# Introduction to
# FIBRE BUNDLES

## LECTURE NOTES
## IN PURE AND APPLIED MATHEMATICS

1. *N. Jacobson,* Exceptional Lie Algebras
2. *L.-Å. Lindahl and F. Poulsen,* Thin Sets in Harmonic Analysis
3. *I. Satake,* Classification Theory of Semi-Simple Algebraic Groups
4. *F. Hirzebruch, W. D. Newmann, and S. S. Koh,* Differentiable Manifolds and Quadratic Forms
5. *I. Chavel,* Riemannian Symmetric Spaces of Rank One
6. *R. B. Burckel,* Characterization of C(X) Among Its Subalgebras
7. *B. R. McDonald, A. R. Magid, and K. C. Smith,* Ring Theory: Proceedings of the Oklahoma Conference
8. *Y.-T. Siu,* Techniques of Extension of Analytic Objects
9. *S. R. Caradus, W. E. Pfaffenberger, and B. Yood,* Calkin Algebras and Algebras of Operators on Banach Spaces
10. *E. O. Roxin, P.-T. Liu, and R. L. Sternberg,* Differential Games and Control Theory
11. *M. Orzech and C. Small,* The Brauer Group of Commutative Rings
12. *S. Thomeier,* Topology and Its Applications
13. *J. M. López and K. A. Ross,* Sidon Sets
14. *W. W. Comfort and S. Negrepontis,* Continuous Pseudometrics
15. *K. McKennon and J. M. Robertson,* Locally Convex Spaces
16. *M. Carmeli and S. Malin,* Representations of the Rotation and Lorentz Groups: An Introduction
17. *G. B. Seligman,* Rational Methods in Lie Algebras
18. *D. G. de Figueiredo,* Functional Analysis: Proceedings of the Brazilian Mathematical Society Symposium
19. *L. Cesari, R. Kannan, and J. D. Schuur,* Nonlinear Functional Analysis of Differential Equations: Proceedings of the Michigan State University Conference
20. *J. J. Schäffer,* Geometry of Spheres in Normed Spaces
21. *K. Yano and M. Kon,* Anti-Invariant Submanifolds
22. *W. V. Vasconcelos,* The Rings of Dimension Two
23. *R. E. Chandler,* Hausdorff Compactifications
24. *S. P. Franklin and B. V. S. Thomas,* Topology: Proceedings of the Memphis State University Conference
25. *S. K. Jain,* Ring Theory: Proceedings of the Ohio University Conference
26. *B. R. McDonald and R. A. Morris,* Ring Theory II: Proceedings of the Second Oklahoma Conference
27. *R. B. Mura and A. Rhemtulla,* Orderable Groups
28. *J. R. Graef,* Stability of Dynamical Systems: Theory and Applications
29. *H.-C. Wang,* Homogeneous Banach Algebras
30. *E. O. Roxin, P.-T. Liu, and R. L. Sternberg,* Differential Games and Control Theory II
31. *R. D. Porter,* Introduction to Fibre Bundles

*Other Volumes in Preparation*

# Introduction to FIBRE BUNDLES

RICHARD D. PORTER

*Northeastern University*
*Boston, Massachusetts*

MARCEL DEKKER, INC. New York and Basel

**Library of Congress Cataloging in Publication Data**

Porter, Richard Dawson.
Introduction to fibre bundles.

(Lecture notes in pure and applied mathematics ; 31)
Bibliography: p.
Includes index.
1. Fiber bundles (Mathematics) I. Title.
QA612.6.P67 516'.362 77-8325
ISBN 0-8247-6626-1



MARCEL DEKKER, INC.
270 Madison Avenue, New York, New York 10016

Current printing (last digit):
10 9 8 7 6 5 4 3 2

PRINTED IN THE UNITED STATES OF AMERICA

# PREFACE

This is a set of lecture notes for a graduate course given at Brown University. The purpose in writing these notes was to bring together some basic material on fibre bundles previously scattered in books and papers.

The first section of Chapter 1 contains the definitions of the classical groups and the definition of homogeneous space. In sections 2 and 3, cellular decompositions of the groups are given and the Pontryagin rings are calculated. The results of sections 2 and 3 are not used until the last section of Chapter 3, so the reader can, if he chooses, proceed from section 1 directly into Chapter 2. The material in Chapter 1 can be found in [Steenrod 1] for an application of the results of Chapter 1 to the existence of vector fields on spheres.

The material in Chapter 2 and in the first three sections of Chapter 3 can be found in [Steenrod 2] and [Husemoller]. The final two sections of Chapter 3 contain a sketch of Milgram's construction of the universal classifying space of a topological group. The reader is referred to [Steenrod 3] and [Milgram] for additional details.

I want to express my appreciation to the Mathematics Department at Brown University for the opportunity to teach the course and for their support in having the notes typed.

Richard D. Porter

CONTENTS

# Introduction to FIBRE BUNDLES

# Chapter 1

# THE CLASSIC GROUPS

## 1. Homogeneous Spaces and the Classic Groups

Definition of Quaternions. The algebra Q of quaternions is the four dimensional vector space over the Reals with basis $\{1,i,j,k\}$ with multiplication defined by setting.

$$i^2 = j^2 = k^2 = -1$$
$$ij = -ji = k$$

by making 1 a two sided unit and by requiring that the multiplication be associative. The conjugate of the quaternion $x = a + ib + jc + kd$ is $\bar{x} = a - jb - jc - kd$ and the norm of a quaternion $x$ is $\bar{x}x$, denoted by $||x||$.

It is easily seen that the quaternions satisfy:

$$ik = -ki = -j$$
$$jk = -kj = i$$
$$||x|| = a^2 + b^2 + c^2 + d^2 \quad \text{so} \quad \bar{x}\cdot x = x\cdot\bar{x}$$
$$||x|| = 0 \quad \text{iff} \quad x = 0$$

if $x \neq 0$, then $\frac{\bar{x}}{||x||}$ is a two sided inverse of $x$

and $||x||\cdot||y|| = ||xy||$.

The Quaternions form an associative (but not commutative) algebra of dimension 4 over the reals. Each nonzero Quaternion has a two sided inverse.

The Quaternions can also be viewed as a two dimensional vector space over the complex numbers $\mathbb{C}$ with basis $\{1,j\}$, multiplication defined by requiring associativity, by requiring that 1 be a two sided unit and by setting $xj = j\bar{x}$ $(x \in \mathbb{C})$. The correspondence between these two points of view is given by $(a+ib) + j(c-id)$ $= a + ib + jc + kd$. If $q = x + jy$, then $\bar{q} = \bar{x} - jy$.

<u>Notation</u>. If $d = 1, 2,$ or $4$, then $F_d = \begin{cases} \text{Reals if } d = 1 \\ \text{Complexes if } d = 2 \\ \text{Quaternions if } d = 4 \end{cases}$

$F$ will be used to denote $F_1, F_2$ or $F_4$. $F^n$ denotes the n-dimensional vector space over $F$ consisting of column vectors with entries in $F$.

<u>Definition</u>. If $x = \begin{pmatrix} x_1 \\ \vdots \\ x_n \end{pmatrix}$ and $y = \begin{pmatrix} y_1 \\ \vdots \\ y_n \end{pmatrix}$ are in $F^n$, the scalar product $\langle x,y \rangle$ is defined by $\langle x,y \rangle = \sum_{i=1}^{n} \bar{x}_i y_i$. The norm of $x$, $||x||$, is defined by $||x|| = \langle x,x \rangle$.

Direct calculation shows that $||x||$ is a real number and $||x|| = 0$ iff $x = 0$. If $||x|| = 1$, then $x$ is called a unit vector in $F^n$. The space of unit vectors in $F^n$ is the sphere, $S^{nd-1}$.

The scalar product satisfies

1. $\langle x, y_1 + y_2 \rangle = \langle x, y_1 \rangle + \langle x, y_2 \rangle$
2. $\langle x_1 + x_2, y \rangle = \langle x_1, y \rangle + \langle x_2, y \rangle$
3. $\langle x, y \rangle = \overline{\langle y, x \rangle}$
4. $\langle x, y\lambda \rangle = \langle x, y \rangle \lambda, \ \lambda \in F$
5. $\langle x\lambda, y \rangle = \bar{\lambda} \langle x, y \rangle, \ \lambda \in F$ using $\overline{(xy)} = \bar{y} \cdot \bar{x}$, $x$ and $y$ in $F$.

<u>Definition</u>. $G(n)$ denotes the set of all linear transformations $L$ of $F^n$ into $F^n$ which preserve scalar products. That is $L \in G(n)$ iff $\langle Lx, Ly \rangle = \langle x, y \rangle$ for all $x$ and $y$ in $F^n$.

It is easily seen that

1. If $L$ and $T$ are in $G(n)$, then $L\,T$ is in $G(n)$.
2. If $L \in G(n)$, then $L$ is invertible and $L^{-1}$ is in $G(n)$.

Thus $G(n)$, with product given by the composition of linear transformations, is a group.

<u>Definition</u>. If $d = 1$, then $G(n)$, denoted by $O(n)$, is called the orthogonal group. If $d = 2$, then $G(n)$ is denoted by $U(n)$ and called the unitary group. If $d = 4$, then $G(n)$ is denoted by $Sp(n)$ and called the symplectic group.

<u>Note</u>: The map $G(1) \to$ (unit sphere in $F$) defined by $L \to L(1)$ is an isomorphism of groups.

So $O(1) = \{-1,1\} = Z_2$

$U(1) =$ unit complexes $= S^1$

$Sp(1) = S^3 =$ unit quaternions

Let $u_i$ be the vector $\begin{pmatrix} 0 \\ \vdots \\ 1 \\ 0 \\ \vdots \\ 0 \end{pmatrix} \leftarrow i^{th}$ place, in $F^n$, then

$\beta = \{u_1 \cdots u_n\}$ is an orthonormal basis for $F^n$. Given a linear transformation $L: F^n \to F^n$, with $L$ in $G(n)$. Set $[L]_{\beta,\beta}$ equal to the $n \times n$ matrix with entries in $F$ whose $i^{th}$ column is the vector $L(u_i)$. The map $L \to [L]_{\beta,\beta}$ defines a group isomorphism between $G(n)$ and a certain subgroup of the group of all invertible $n \times n$ matrices with entries in $F$.

Proposition 1.1.1. $G(n)$ is isomorphic to the group of all $n \times n$ matrices, $A$, over $F$ with the property that the columns of $A$ form an orthonormal basis for $F^n$, which in turn equals the group of all $n \times n$ matrices over $F$, with the property that $\overline{A}^t A = I$.

Proof. Let $L \in G(n)$. $L$ preserves scalar products and so $||L(x)|| = ||x||$ for all $x$ in $F^n$. Thus $L$ takes any orthonormal basis for $F^n$ into an orthonormal basis for $F^n$. In particular, $\{Lu_1 \cdots Lu_n\}$ is an orthonormal basis for $F^n$. So if $L \in G(n)$, then the columns of $[L]_{\beta,\beta}$ form an orthonormal basis for $F^n$.

Conversely, let $A$ be any $n \times n$ matrix with entries in $F$ whose columns form an orthonormal basis for $F^n$. Since $Au_i = i^{th}$ column of $A$, we have that $\langle Au_i, Au_j \rangle = \delta_{ij}$ where

$$\delta_{ij} = \left\{ \begin{matrix} 0 & \text{if} & i \neq j \\ 1 & \text{if} & i = j \end{matrix} \right\} .$$

Let $x$ and $y$ be in $F^n$, to show $\langle Ax, Ay \rangle = \langle x, y \rangle$, write $x = \sum_{i=1}^{n} u_i x_i$, $y = \sum_{i=1}^{n} u_i y_i$ with $x_i$ and $y_i$ in $F$. Then

$$\begin{aligned} \langle Ax, Ay \rangle &= \langle A\Sigma u_i x_i, A\Sigma u_j y_j \rangle \\ &= \langle \Sigma A u_i x_i, \Sigma A u_j y_j \rangle \\ &= \sum_i \sum_j \langle A u_i x_i, A u_j y_j \rangle \\ &= \sum_i \sum_j \bar{x}_i \langle A u_i, A u_j \rangle y_i \\ &= \sum_i \sum_j \bar{x}_i \delta_{ij} y_j = \sum_i \bar{x}_i y_i = \langle x, y \rangle . \end{aligned}$$

This, together with $[T \circ L]_{\beta,\beta} = [T]_{\beta,\beta} \circ [L]_{\beta,\beta}$, proves the first part of Proposition 1.1.1.

The second part follows from the observation that the $i,j$ entry of the matrix $\bar{A}^t A$ is the scalar product $\langle Au_i, Au_j \rangle$. From now on $G(n)$ will be identified with the corresponding group of matrices.

Definition. If G is a group and a topological space, then G is called a topological group if

1. $G \to G$ defined by $x \to x^{-1}$ is continuous

and

2. $G \times G \to G$ defined by $(x,y) \to xy$ is continuous.

Examples of topological groups.

1. $G(n)$ topologized as a subspace of $F^{n^2}$.
2. $R^1$ and $R^n$ (additively)
3. $R^1 - \{0\}$ (multiplicatively)
4. For $d = 1$ or 2 set $GL(n,F_d)$ equal to the group of all invertible $n \times n$ matrices with entries in $F_d$. Topologize $GL(n,F_d)$ as a subspace of $F_d^{n^2}$. For $d = 1$ or 2 set $SL(n,F_d)$ equal to the subgroup of $GL(n,F_d)$ consisting of matrices with determinant 1. Set $SO(n)$ equal to the subgroup of $O(n)$ consisting of matrices with determinant 1, and set $SU(n)$ equal to the subgroup of $U(n)$ consisting of matrices with determinant 1.
5. Let G be a group, then G with discrete topology is a topological group.
6. A product of topological groups is a topological group.
7. If H is a closed, normal subgroup of G, then G/H with the quotient space topology is a topological group. For example if $I^n$ denotes the subgroup of $R^n$ consisting of vectors with integer coordinates, then $R^n/I^n$ can be

identified with the product $\underbrace{S' \times \cdots \times S'}_{n}$ of topological groups.

8. If $X$ is a compact, Hausdorff space; set $h(X)$ equal to the group of homeomorphisms of $X$ onto itself. Then $h(X)$ with the compact open topology is a topological group.

<u>Definition</u>. Let $G$ be a topological group, and let $H$ be a closed subgroup, then the space of left cosets, $G/H$, topologized as a quotient space of $G$ is called a homogeneous space.

<u>Examples of homogeneous spaces</u>.

1. If $k < n$, define $G(k) \to G(n)$ by $A \to \begin{pmatrix} A & 0 \\ 0 & 1 \end{pmatrix}$. In this way $G(k)$ is a closed subgroup of $G(n)$. The homogeneous space $G(n)/G(k)$, denoted by $G(n,k)$, is the Stiefel manifold of ordered orthonormal $(n-k)$ frames in $F^n$. For example $G(n,n-1)$ is homeomorphic to the sphere $S^{nd-1}$ for $n \geq 2$.

2. If $k < n$, define $G(k) \times G(n-k) \to G(n)$ by $(A,B) \to \begin{pmatrix} A & 0 \\ 0 & B \end{pmatrix}$. In this way, $G(k) \times G(n-k)$ is a closed subgroup of $G(n)$. The homogeneous space $G(n)/G(k) \times G(n-k)$, denoted by $M(n,k)$ is the Grassmanian manifold of $(n-k)$ dimensional subspaces of $F^n$. Thus $M(n,k)$ is homeomorphic to $M(n,n-k)$ and $M(n,n-1)$ is the real, complex, Quaternionic projective space of real dimension $d(n-1)$.

To see that $G(n,k)$ is the space of ordered orthonormal $(n-k)$

frames in $F^n$, it suffices to prove that $AG(k) = BG(k)$ iff $A$ and $B$ have the same last $(n-k)$ columns. This is done by observing that if $C \varepsilon G(k)$, $A \varepsilon G(n)$; then $AC$ and $A$ have the same last $(n-k)$ columns, and that if $A$ and $B$ have the same last $(n-k)$ columns, then $A^{-1}B \varepsilon G(k)$.

To see that $M(n,k)$ is the space of $(n-k)$ dimensional subspaces of $F^n$, it suffices to prove that $A(G(k) \times G(n-k))$ equals $B(G(k) \times G(n-k))$ iff the last $(n-k)$ columns of $A$ and the last $(n-k)$ columns of $B$ are bases for the same $(n-k)$ dimensional subspaces of $F^n$. This is done by observing that if $C \varepsilon G(k) \times G(n-k)$, $A \varepsilon G(n)$; then the last $(n-k)$ columns of $AC$ and the last $(n-k)$ columns of $A$ span the same subspace of $F^n$, and that if the last $(n-k)$ columns of $A$ and the last $(n-k)$ columns of $B$ are bases for the same subspaces of $F^n$, then $A^{-1}B \varepsilon G(k) \times G(n-k)$. Note that since $A^{-1} = \overline{A}^t$, the $(i,j)^{th}$ entry of the matrix $A^{-1}B$ is the scalar product $\langle i^{th}$ column of $A$, $j^{th}$ column of $B\rangle$.

## 2. Cell Structure of $O(n), U(n), Sp(n)$

Definition of $\phi$. Let $S^{nd-1}$ be the sphere of unit vectors in $F^n$. Define $\phi: S^{nd-1} \times S^{d-1} \to G(n)$ by requiring that $\phi(x,\lambda)y = y$ if $\langle x,y\rangle = 0$ and $\phi(x,\lambda)x = x\lambda$. Then

$$\phi(x,\lambda)y = x(\lambda-1)\langle x,y\rangle + y$$

$$[\phi(x,\lambda)]_{ij} = x_i(\lambda-1)\overline{x}_j + \delta_{ij} \quad \text{in matrix notation.}$$

If $m < n$, the usual inclusion of $F^m$ in $F^n$ induces a commutative diagram

$$\begin{array}{ccc} S^{md-1} \times S^{d-1} & \longrightarrow & S^{nd-1} \times S^{d-1} \\ \downarrow \phi & & \downarrow \phi \\ G(m) & \longrightarrow & G(n) \end{array}$$

<u>Definition of $Q_n$.</u> Let $Q_n$ be the quotient space of $S^{nd-1} \times S^{d-1}$ induced by $\phi$ for $n \geq 1$, and let $Q_o$ be a single point. Map $Q_o$ into $G(n)$ by $Q_o \to I$.

<u>Note</u>: 1. The diagram $\begin{array}{ccc} Q_m & \longrightarrow & Q_n \\ \downarrow & & \downarrow \\ G(m) & \longrightarrow & G(n) \end{array}$ is commutative.

2. $Q_n$ is the quotient space of $S^{nd-1} \times S^{d-1}$ by the relations $(x,\lambda) = (xv, v^{-1}\lambda v)$ and $(x,1) = (y,1)$.

A direct calculation shows that $\phi(x,\lambda) = \phi(xv, v^{-1}\lambda v)$ and $\phi(x,\lambda) = I$ iff $\lambda = 1$.

To see that these are all the relations, suppose $\phi(x,\lambda) = \phi(y,\gamma) \neq I$. Considering the fixed point sets of $\phi(x,\lambda)$ and $\phi(y,\gamma)$ shows that $y = xv$ for some $v$ in $S^{d-1}$.

$\phi(x,\lambda)x = x\lambda$ so $\phi(x,\lambda)xv = x\lambda v$ and $\phi(xv,\gamma) = xv\gamma$. Thus $\phi(x,\lambda) = \phi(xv,\gamma)$ implies $v\gamma = \lambda v$ so $\gamma = v^{-1}\lambda v$.

3. $Q_n$ is a compact Hausdorff space, $Q_n \to G(n)$ and

$Q_m \to Q_n$ for $m \leq n$, are embeddings

$Q_n$ is compact since it is the image of $S^{nd-1} \times S^{d-1}$ which is compact. Since $Q_n$ is compact and $Q_n \to G(n)$ is 1-1, the map $Q_n \to G(n)$ is an embedding. (Recall any continuous 1-1 map of a compact space into a Hausdorff space is an embedding.)

The next step is to show that $Q_n$ is a C. W. complex.

Consider $E^{(n-1)d}$ as the ball of vectors $x$ in $S^{nd-1} \subseteq F^n$ with $x_n$ real and $\geq 0$. (Note: $x_n = \sqrt{1 - \sum_{i=1}^{n-1} \bar{x}_i x_i}$.) Let $f_n\colon E^{(n-1)d} \to S^{nd-1}$ be the inclusion map. Let $g\colon (E^{d-1}, S^{d-2}) \to (S^{d-1}, 1)$ be the usual relative homeomorphism ($S^{-1} = \phi$, $E^0 = \{-1\}$). Set $h_n\colon E^{nd-1} \to Q_n (n \geq 1)$ equal to the composition

$$E^{nd-1} = E^{(n-1)d} \times E^{d-1} \xrightarrow{f_n \times g} S^{nd-1} \times S^{d-1} \xrightarrow{\phi} Q_n.$$

<u>Lemma 1.2.1</u>. The map $h_n$ defines a relative homeomorphism $h_n\colon (E^{nd-1}, S^{nd-2}) \to (Q_n, Q_{n-1})$, if $n \geq 1$. Therefore $Q_n$ is a C. W. Complex with a 0-cell $Q_0$ and with an $(md-1)$-cell for each $m$ such that $1 \leq m \leq n$.

<u>Proof</u>. The image of $E^{nd-1} - S^{nd-2}$ in $S^{nd-1} \times S^{d-1}$ equals $\{(x,\lambda)\colon x_n$ is real and $> 0, \lambda \neq 1\}$, and $Q_n - Q_{n-1} = \{\phi(x,\lambda)\colon x_n \neq 0$ and $\lambda \neq 1\}$. So $h_n$ maps $E^{nd-1} - S^{nd-2}$ into $Q_n - Q_{n-1}$ and is

1-1 since $x_n v$ and $x_n$ both real and $> 0$ with $v \varepsilon S^{d-1}$ implies $v = 1$. Let $\phi(x,\lambda) \varepsilon Q_n - Q_{n-1}$. Since $x_n \neq 0$, there is a $v$ in $S^{d-1}$ such that $x_n v$ is real and $> 0$ (set $v = \bar{x}_n/\sqrt{||x_n||}$). Then $\phi(x,\lambda) = \phi(xv, v^{-1}\lambda v)$ with $(xv, v^{-1}\lambda v)$ in the image of $E^{nd-1} - S^{nd-2}$.

Thus $h_n\colon E^{nd-1} - S^{nd-2} \to Q_n - Q_{n-1}$ is 1-1 and onto. Since $E^{nd-1}$ is compact and $Q_n$ is Hausdorff $h_n\colon (E^{nd-1}, S^{nd-2}) \to (Q_n, Q_{n-1})$ is a relative homeomorphism. The rest of the lemma follows by induction on $n$.

<u>Definition</u>. Let $u$ be the multiplication in $G(n)$. A <u>normal cell</u> of $G(n)$ is a map (or the image of a map) of the form

$$E^{i_1 d-1} \times E^{i_2 d-1} \times \cdots \times E^{i_r d-1} \xrightarrow{h\times h\times\cdots\times h} Q_{i_1} \times Q_{i_2} \times \cdots \times Q_{i_r} \xrightarrow{u} G(n)$$

where $n \geq i_1 > i_2 > i_3 > \cdots > i_r > 0$. We denote such a cell by $(i_1 \cdots i_r|n)$ or simply by $(i_1 \cdots i_r)$ if this will cause no confusion.

Our next objective is to prove: $G(n)$ is a C. W. complex whose cells are the normal cells and the 0-cell $I$. The map $u\colon Q_n \times G(n-1) \to G(n)$ is cellular and induces an epimorphism of chain complexes.

Let $\alpha\colon Q_n \to S^{nd-1}$ be the composition $Q_n \to G(n) \xrightarrow[\text{column}]{\text{Proj to last}} S^{nd-1}$.

Lemma 1.2.2. The map $\alpha: (Q_n, Q_{n-1}) \to (S^{nd-1}, u_n)$ is a relative homeomorphism.

Proof. First observe that $\alpha$ maps $Q_n - Q_{n-1}$ into $S^{nd-1} - u_n$. For if $x_n(\lambda-1)\bar{x}_n + 1 = 1$ with $x_n \neq 0$ then $\lambda = 1$. Recall $Q_n - Q_{n-1} = \{\phi(x,\lambda): x_n$ real and $> 0, \lambda \neq 1\}$ and

$$\alpha\phi(x,\lambda) = \begin{pmatrix} x_1(\lambda-1)\bar{x}_n \\ \cdot \\ \cdot \\ \cdot \\ x_{n-1}(\lambda-1)\bar{x}_n \\ x_n(\lambda-1)\bar{x}_n + 1 \end{pmatrix}.$$

Since $Q_n$ is compact and $S^{nd-1}$ is Hausdorff it suffices to prove that $\alpha: Q_n - Q_{n-1} \to S^{nd-1} - u_n$ is 1-1 and onto.

Let $y \in S^{nd-1} - u_n$. We want to find $(x,\lambda) \in S^{nd-1} \times S^{d-1}$ such that $x_n$ is real and $> 0$, and $x(\lambda-1)x_n = y - u_n$, showing that the choice of $(x,\lambda)$ is unique.

In the real d-dimensional space F, $(y_n-1)$ lies in the closed ball bounded by the sphere of scalars of the form $(\lambda-1)$ where $|\lambda| = 1$.

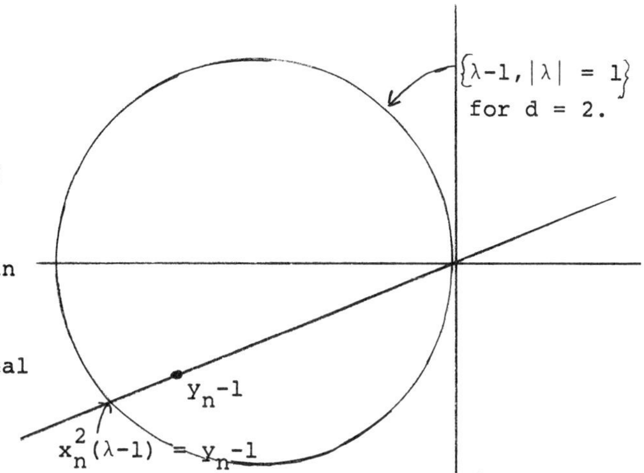


Diagram 1.

Note $y_n - 1 \neq 0$. Projecting from the origin in the real d-dimensional space F, we can solve uniquely the equation $x_n^2(\lambda-1) = y_n - 1$ for $x_n$ real and $> 0$, $|\lambda| = 1$, $\lambda \neq 1$. Knowing $\lambda$ and $x_n$, $x_i$ is determined uniquely for $1 \leq i \leq n-1$. We now have $x(\lambda-1)x_n = y - u_n$, $x_n$ real, $|\lambda| = 1$, and $\lambda \neq 1$.

It remains to show that $\langle x,x\rangle = 1$.

$$\langle x(\lambda-1)x_n, x(\lambda-1)x_n\rangle = \langle y-u_n, y-u_n\rangle$$

$$(\bar{\lambda}-1)x_n\langle x, x(\lambda-1)x_n\rangle = \langle y,y\rangle + \langle y,-u_n\rangle + \langle -u_n,y\rangle + \langle -u_n,-u_n\rangle$$

$$(\bar{\lambda}-1)x_n\langle x,x\rangle(\lambda-1)x_n = 1 - \bar{y}_n - y_n + 1$$

$$\langle x,x\rangle(\bar{\lambda}-1)(\lambda-1)x_n^2 = 2 - \bar{y}_n - y_n$$

$$\langle x,x\rangle(1-\bar{\lambda}-\lambda+1)x_n^2 = 2 - \bar{y}_n - y_n$$

$$\langle x,x\rangle(2-\bar{\lambda}-\lambda)x_n^2 = 2 - \bar{y}_n - yn$$

Thus it suffices to prove $(2-\bar{\lambda}-\lambda)x_n^2 = 2 - \bar{y}_n - y_n \neq 0$.

We have $x_n^2(\lambda-1) = y_n - 1$ or equivalently $(1-\lambda)x_n^2 = 1 - y_n$

so $x_n^2(\bar{\lambda}-1) = \bar{y}_n - 1$ $\qquad (1-\bar{\lambda})x_n^2 = 1 - \bar{y}_n$

Thus

$$(1-\lambda)x_n^2 + (1-\bar{\lambda})x_n^2 = 1 - y_n + 1 - \bar{y}_n$$

so $(2-\lambda-\bar{\lambda})x_n^2 = 2 - y_n - \bar{y}_n$. If $2 - \lambda - \bar{\lambda} = 0$ then $\lambda + \bar{\lambda} = 2$ so $2 \operatorname{Re} \lambda = 2$ or $\operatorname{Re} \lambda = 1$. But $|\lambda| = 1$ and $\lambda \neq 1$. Thus $2 - \lambda - \bar{\lambda} \neq 0$. So we have

$$\langle x,x\rangle(2-\bar{\lambda}-\lambda)x_n^2 = (2-\bar{\lambda}-\lambda) \text{ with } x_n \text{ real} > 0. \quad 2 - \bar{\lambda} - \lambda \neq 0.$$

Thus $\langle x,x\rangle = 1$. Which proves the Lemma.

Lemma 1.2.3. If $n \geq 1$, $u:(Q_n \times G(n-1), Q_{n-1} \times G(n-1)) \to (G(n),G(n-1))$ is a relative homeomorphism and maps $Q_n \times G(n-1)$ onto $G(n)$.

Proof. Note $Q_n \times G(n-1) - Q_{n-1} \times G(n-1) = (Q_n-Q_{n-1}) \times G(n-1)$. As usual it suffices to show that $u$ induces a 1-1 onto map of $(Q_n-Q_{n-1}) \times G(n-1)$ into $G(n) - G(n-1)$.

(1) $u$ maps $(Q_n-Q_{n-1}) \times G(n-1)$ into $G(n) - G(n-1)$

For suppose $A \in Q_n - Q_{n-1}$, $B \in G(n-1)$ and $AB \in G(n-1)$. Then $A \in G(n-1)$. But Lemma 1.2.2 shows that $(Q_n-Q_{n-1}) \cap G(n-1) = \phi$. So $AB \notin G(n-1)$ if $A \in Q_n - Q_{n-1}$ and $B \in G(n-1)$.

(2) $u: (Q_n-Q_{n-1}) \times G(n-1) \to G(n) - G(n-1)$ is 1-1 and onto

first 1-1. Let $A,A^1 \in Q_n - Q_{n-1}$, $B,B^1 \in G(n-1)$ with $AB = A^1B^1$ then $A$ and $A^1$ have the same last column which by Lemma 2.2 is $\neq u_n$. From Lemma 1.2.2 we have $A = A^1$ so $B = B^1$ which proves 1-1.

> Next onto. Let $A \in G(n) - G(n-1)$. Then last column of $A \neq u_n$ so there is an $A^1$ in $Q_n - Q_{n-1}$ so that A and $A^1$ have the same last column. Thus $AG(n-1) = A^1G(n-1)$ i.e. $A \in A^1G(n-1)$, which proves onto.

At first glance it looks as though Lemma 1.2.3 and induction is enough to show that $G(n)$ is a C. W. complex whose cells are the normal cells and the 0-cell I, and that $u: Q_n \times G(n-1) \to G(n)$ is cellular. The proposed proof breaks down because we can not show $u: Q_n \times G(n) \to G(n)$ is cellular.

Lemma 1.2.4. Let $x \in S^{nd-1} \subset F^n$, let $\lambda \in S^{d-1} \subset F$ and let $A \in G(n)$. Then $A\phi(x,\lambda)A^{-1} = \phi(Ax,\lambda)$.

Proof. Show

1. $A\phi(x,\lambda)A^{-1}y = y$ if $\langle Ax,y\rangle = 0$
   (if $\langle Ax,y\rangle = 0$ then $\langle x,A^{-1}y\rangle = 0$).
2. $A\phi(x,\lambda)A^{-1}x = x\lambda$.

Lemma 1.2.5. $u: (Q_n \times Q_n) = u(Q_n \times Q_{n-1}) \subset G(n)$ for $n \geq 1$.

Proof. We shall reduce the case where n is arbitrary to the case where $n = 2$. We therefore begin by checking the lemma for $n = 1$ and $n = 2$.

If $n = 1$ then $Q_1 = G(1)$. To see this recall $Q_n$ is the image of $S^{nd-1} \times S^{d-1}$ in $G(n)$ under the map $[\phi(x,\lambda)]_{i,j} = x_i(\lambda-1)\bar{x}_j + \delta_{ij}$.

Thus $Q_1$ is the image of $S^{d-1} \times S^{d-1}$ in $S^{d-1}$ under the map $(x,\lambda) \to x(\lambda-1)\bar{x} + 1 = x\lambda\bar{x}$. Since $(1,\lambda) \to \lambda$, $\phi: S^{d-1} \times S^{d-1} \to S^{d-1} = G(1)$ is onto. So $Q_1 = G(1)$. Recall $Q_o = id$. Thus $u(Q_1 \times Q_1) = G(1) = u(Q_1 \times Q_o)$.

If $n = 2$ then $u(Q_2 \times Q_1) = G(2)$. To see this observe $Q_1 = G(1)$ from above so $u(Q_2 \times Q_1) = u(Q_2 \times G(1))$. But $u(Q_2 \times G(1)) = G(2)$ by

Lemma 1.2.3. So

$$G(2) = u(Q_2 \times Q_1) \subseteq u(Q_2 \times Q_2) \subseteq G(2).$$

Now let $(x,y) \in S^{nd-1}$ and let $\lambda, v \in S^{d-1}$ be arbitrary. We want to show that $\phi(x,\lambda)\phi(y,v)$ is in $Q_nQ_{n-1}$.

Pick a matrix $A$ in $G(n)$ such that $Ax$, and $Ay$ are in $F^2$ and $A^{-1}F^1 \subseteq F^{n-1}$. Let $Ax = x^1$, and $Ay = y^1$ then

$$\begin{aligned} \phi(x,\lambda)\phi(y,v) &= \phi(A^{-1}x^1,\lambda)\phi(A^{-1}y^1,v) \\ &= A^{-1}\phi(x^1,\lambda)AA^{-1}\phi(y^1,v)A \\ &= A^{-1}\phi(x^1,\lambda)\phi(y^1,v)A \end{aligned}$$

$\phi(x^1,\lambda)\phi(y^1,v) \in Q_2Q_2 = Q_2Q_1$ so there are $(x_1,\lambda_1)$ in $S^{2d-1} \times S^{d-1}$ and $(y_1,v_1)$ in $S^{d-1} \times S^{d-1}$ such that $\phi(x^1,\lambda)\phi(y^1,v) = \phi(x_1,\lambda_1)\phi(y_1,v_1)$. So

$$\begin{aligned}\phi(x,\lambda)\phi(y,v) &= \phi(A^{-1}x^1,\lambda)\phi(A^{-1}y^1,v)\\ &= A^{-1}\phi(x_1,\lambda_1)\phi(y_1,v_1)A\\ &= A^{-1}\phi(x_1,\lambda_1)AA^{-1}\phi(y_1,v_1)A\\ &= \phi(A^{-1}x_1,\lambda_1)\phi(A^{-1}y_1,v_1)\\ &\varepsilon\ Q_nQ_{n-1} \quad \text{since} \quad A^{-1}y_1 \ \varepsilon\ F^{n-1}.\end{aligned}$$

This completes the proof of Lemma 1.2.5.

We can now prove

Proposition 1.2.6.

a)n $G(n)$ is a C.W. complex, whose cells are the normal cells and the 0-cell $I$. The map

$u: Q_n \times G(n-1) \to G(n)$ is cellular and induces an epimorphism of chain complexes.

b)n Let $n \geq i_1 \cdots i_r > 0$. Then $Q_{i_1}Q_{i_2} \cdots Q_{i_r}$ is contained in the $\sum_{s=1}^{r}(i_sd-1)$ skeleton of $G(n)$.

c)n $u: Q_n \times G(n) \to G(n)$ is cellular.

Proof. Recall $G(0) = id$, $G(1) = S^{d-1} = Q_1$. a,b, and c(n) are easily seen to be true for $n = 1$. The proof proceeds by induction using the previous lemmas. We shall use, without further comment, the fact that $Q_n$ is a C. W. complex with 0-cell 1 and an $md - 1$ cell, denoted by $(m)$, for each $1 \leq m \leq n$, and that $(m)$

maps to the normal cell (m) in $G(n)$. We state the additional lemmas to be used in the proof.

<u>Lemma 1.2.3</u>. If $n \geq 1$, $u: (Q_n \times G(n-1), Q_{n-1} \times G(n-1)) \to (G(n), G(n-1))$ is a relative homeomorphism. Moreover, u maps $Q_n \times G(n-1)$ onto $G(n)$.

<u>Lemma 1.2.4</u>. $A\phi(x,\lambda)A^{-1} = \phi(Ax,\lambda)$.

<u>Lemma 1.2.5</u>. $u: (Q_n \times Q_n) = u(Q_n \times Q_{n-1}) \subseteq G(n)$ for $n \geq 1$.

Assume a,b, and c (n-1) are true.

(1) a(n) is true.

By Lemma 1.2.3 $u(Q_n \times G(n-1), Q_{n-1} \times G(n-1) \to (G(n),G(n-1))$ is a relative homeomorphism. Using a(n-1) we see that $G(n)$ is obtained by adjoining to $G(n-1)$ the cells $(m) \times (i_1 \cdots i_r)$, $n - 1 \geq i_1 > i_2 \cdots > i_r > 0$ and the cell $(n) \times 1$.

From c(n-1), $u: Q_{n-1} \times G(n-1) \to G(n-1)$ is cellular and a(n) follows.

(2) b(n) is true. First observe that $Q_tQ_j \subseteq Q_jQ_t$ in $G(n)$ for $t \leq j \leq n$.

Let $x \in S^{td-1}$, $y \in S^{jd-1}$ and consider $\phi(x,\lambda)\phi(y,v) \in Q_tQ_j$.

Set $A = \phi(x,\lambda)$, then from Lemma 1.2.4 we have $\phi(x,\lambda)\phi(y,v) = A\phi(y,v) = \phi(Ay,v)A = \phi(Ay,v)\phi(x,\lambda)$, $A \in G(t) \subseteq G(j)$ and $y \in S^{jd-1}$ so $Ay \in S^{jd-1}$ and hence $\phi(Ay,v) \in Q_j$.

Thus $\phi(x,\lambda)\phi(y,v) = \phi(Ay,v)\phi(x,\lambda) \in Q_jQ_t$ so $Q_tQ_j \subseteq Q_jQ_t$ if $t \leq j$.

From Lemma 1.2.5 and the result that $Q_tQ_j \subseteq Q_jQ_t$ if $t \leq j$, we can assume $n \geq i_1 > i_2 > i_3 > \cdots > i_r > 0$ in the hypothesis of b(n). From b(n-1) it follows that $Q_{i_2} \cdots Q_{i_r}$ is contained in the $\sum_{i=2}^{r}(i_sd-1)$ skeleton of G(n-1). The desired result follows since $Q_{i_1}$ has dimension $i_1d-1$ and since $u: Q_{i_1} \times G(n-1) \to G(n)$ is cellular for $i_1 \leq n$ by a(n). Thus b(n) is true.

(3) c(n) is true. Since $Q_o$ is the identity, u is cellular on $Q_o \times G(n)$. By a(n), u is cellular on $Q_o \times G(n-1)$. So we have only to check that u is cellular on cells of the form $(t) \times (n,i_1 \cdots i_r)$ where $n \geq t > 0$, $n > i_1 > i_2 \cdots > i_r > 0$. $u((t) \times (n,i_1 \cdots i_r)) \subseteq Q_tQ_nQ_{i_1} \cdots Q_{i_r}$ which is contained in the $(td-1) + (nd-1) + \sum_{s=1}^{r}(i_sd-1)$ skeleton of G(n) by b(n).

This completes the proof of Proposition 1.2.6.

## 3. Pontryagin Rings of the Groups G(n)

Lemma 1.3.1. If $d = 1$, $Q_n$ is the disjoint union of the point $Q_o$ and $RP^{n-1}$. The embedding of $Q_{n-1}$ in $Q_n$ $(n \geq 2)$ corresponds to the usual embedding of $RP^{n-2}$ in $RP^{n-1}$ and $Q_n - Q_o$ consists entirely of matrices with determinant $-1$.

Proof. Recall $Q_n$ is the quotient of $S^{nd-1} \times S^{d-1}$ by the relations $(x,1) = (y,1)$ and $(x,\lambda) = (xv, v^{-1}\lambda v)$. So for $d = 1$, $Q_n$ is the quotient space of $S^{n-1} \times S^o$ by the relations $(x,1) = (y,1)$ and $(x,-1) = (-x,-1)$. It follows that $Q_n$ is the disjoint union of $RP^{n-1}$ and $Q_o$, and the embedding of $Q_{n-1}$ in $Q_n$ $(n \geq 2)$ corresponds to the usual embedding of $RP^{n-2}$ in $RP^{n-1}$.

If $A \in O(n)$, then $A^{-1} = A^t$ and since $\det A = \det A^t$ we have $1 = \det(I) = \det(AA^t) = \det(A)\det(A^t) = [\det(A)]^2$. So all matrices in $O(n)$ have determinant $\pm 1$. For $d = 1$, $Q_1 - Q_o$ consists of the single $1 \times 1$ matrix $(-1)$. For $n \geq 1$, $d = 1$, $Q_n - Q_o$ is connected, and contains the point $Q_1 - Q_o$, and $\det\colon O(n) \to S^o$ is continuous so every matrix in $Q_n - Q_o$ must have determinant $-1$.

For the remainder of this chapter, all chain complexes and all homology groups will be taken with coefficients $R$, where $R$ is a commutative ring with unit if $d = 2$, or $4$ and $R = Z_2$ = integers mod 2 if $d = 1$.

It follows from Lemma 1.3.1 that if $d = 1$, the boundary of each cell in $Q_n$ is algebraically zero. If $d = 2$ or $4$, the boundary of each cell in $Q_n$ is $0$ for dimensional reasons. Using induction (recall $Q_1 = G(1)$) and the fact that $u\colon Q_n \times G(n-1) \to G(n)$ is an epimorphism of chain complexes we see that the boundary operator on the cells of $G(n)$ is identically 0.

Definition. If $(i_1 \cdots i_r)$ is a normal cell of $G(n)$ then $[i_1 \cdots i_r]$ denotes its homology class (called a normal class); $\underline{1}$ denotes the homology class of the 0-cell $I$.

From Proposition 1.2.6 and the discussion above we have

Lemma 1.3.2. $H_*(G(n),R)$ is the free R-module generated by $\underline{1}$ and the normal classes $[i_1 \cdots i_r]$. The inclusion map $G(m) \to G(n)$ for $m \leq n$ induces the obvious map on homology. The map $u_*\colon H_*(Q_n \times G(n-1);R) \to H_*(G(n);R)$ is an epimorphism.

Since $G(n)$ is a topological group, the map $H_*(G(n):R) \otimes H_*(G(n);R) \to H_*(G(n);R)$ induced by $u_*$ gives $H_*(G(n);R)$ the structure of an associative algebra over $R$ with unit, called the Pontryagin Ring of $G(n)$.

Theorem 1.3.3. The Pontryagin Ring of $G(n)$ is the associative algebra over $R$ with unit $\underline{1}$, generated by the normal classes $[i]$ of dimension $(id-1)$ where $n \geq i > 0$ subject to the relations

| $d = 1$ | $d = 2$ or $4$ |
|---|---|
| $[1][1] = \underline{1}$ | |
| $[i][i] = 0$ if $i > 1$ | $[i][i] = 0$ for all $i$ |
| $[i][j] = [j][i]$ if $i \neq j$ | $[i][j] = -[j][i]$ if $i \neq j$ |

Proof. First we show that the above relations hold. Consider the case $d = 1$. $[1][1] = \underline{1}$ follows since $[1]$ is the homology class of the 0-cell $\begin{pmatrix} -1 & 0 \\ 0 & I \end{pmatrix}$ and $\begin{pmatrix} -1 & 0 \\ 0 & I \end{pmatrix}\begin{pmatrix} -1 & 0 \\ 0 & I \end{pmatrix} = I$. Assume $i > 1$. $\dim [i] = i - 1$, so $\dim([i][i]) = 2i - 2$. Now since the normal cell $(i)$ is contained in $Q_i$ and since $Q_iQ_i \subset Q_iQ_{i-1} \subset (i-1) + (i-1) - 1 = 2i - 3$ skeleton of $0(n)$ (for $i > 1$) we see that $[i][i] = 0$ for dimensional reasons. A similar argument shows that $[i][i] = 0$ for all $i$ if $d = 2$, or $4$.

Next we show that $[i][j] = [j][i]$ if $i \neq j$ and $d = 1$
and $[i][j] = -[j][i]$ if $i \neq j$ and $d = 2$ or $4$.

Suppose we can find a map

$$f\colon E^{id-1} \times E^{jd-1} \to E^{jd-1} \times E^{id-1}$$

which is a map of pairs

$$f\colon (E^{id-1}, S^{id-2}) \times (E^{jd-1}, S^{jd-2}) \to (E^{jd-1}, S^{jd-2}) \times (E^{id-1}, S^{id-2})$$

such that

$$\begin{array}{ccccc} E^{id-1} \times E^{jd-1} & & \xrightarrow{\quad f \quad} & & E^{jd-1} \times E^{id-1} \\ \Big\downarrow h_i \times h_j & & & & \Big\downarrow h_j \times h_i \\ Q_i \times Q_j & \xrightarrow{\ u\ } & G(n) & \xleftarrow{\ u\ } & Q_j \times Q_i \end{array}$$

commutes.

<u>Notation</u>: $H_{jd-1}(Q_j, Q_{j-1})$ is isomorphic to $Z$ and the class determined by the cell $(j)$ is a generator. Denote this class by $[j]$. Let $[j]$ also denote the class in $H_{jd-1}(E^{jd-1}, S^{jd-2})$ which

is mapped to the class $[j]$ in $H_{jd-1}(Q_j,Q_{j-1})$ by the isomorphism $(h_j)_*\colon H_{jd-1}(E^{id-1},S^{jd-2}) \to H_{jd-1}(Q_j,Q_{j-1})$. $H_{(i+j)d-2}((E^{id-1},S^{id-2}) \times (E^{jd-1},S^{jd-2});R) \cong R$ and is generated by the cross product $[i] \times [j]$. Applying $H_{(i+j)d-2}(\ ;R)$ to the commutative diagram of pairs.

$$\begin{array}{ccccc}
(E^{id-1},S^{id-2})\times(E^{jd-1},S^{jd-2}) & & \longrightarrow & & (E^{jd-1},S^{jd-2})\times(E^{id-1},S^{id-2}) \\
\downarrow h_i\times h_j & & & & \downarrow h_j\times h_i \\
(Q_i,Q_{i-1})\times(Q_j,Q_{j-1}) & \longrightarrow & (G(n),(i+j)d-2-d \text{ skeleton}) & \longleftarrow & (Q_j,Q_{j-1})\times(Q_i,Q_{i-1})
\end{array}$$

we get

$$\begin{array}{ccccc}
[i]\times[j] & & \longrightarrow & & \alpha([j]\times[i]) \\
\downarrow & & & & \downarrow \\
[i]\times[j] & \longrightarrow & [i][j] = \alpha[j][i] & \longleftarrow & \alpha([j]\times[i])
\end{array}$$

Thus $[i][j] = \alpha[j][i]$ in $H_{(i+j)d-2}((G(n)\colon (i+j)d-2-d \text{ skeleton});R)$. $(i+j)d-2-d \leq (i+j)d-2-1$ and boundaries in $G(n)$ are $0$ so $H_{(i+j)d-2}((G(n),(i+j)d-2-d \text{ skeleton});R) \cong H_{(i+j)d-2}(G(n);R)$ and hence $[i][j] = \alpha[j][i]$ in $H_*(G(n);R)$. Thus to show

$$[i][j] = [j][i] \quad d = 1,\ i \neq j$$

and

$$[i][j] = -[j][i] \quad d = 2 \text{ or } 4$$

it suffices to find a suitable map $f$ and then calculate $f_*$.

Let $1 \leq j < i$ and $d = 1,2$ or $4$. Then the diagram

$$\begin{array}{ccccc}
E^{id-1} \times E^{jd-1} & \xrightarrow{\psi} & E^{jd-1} \times E^{id-1} & \xrightarrow{\theta} & E^{jd-1} \times E^{id-1} \\
\downarrow h_i \times h_j & & & & \downarrow h_j \times h_i \\
Q_i \times Q_j & \xrightarrow{\mu} & G(n) & \xleftarrow{u} & Q_j \times Q_i
\end{array}$$

is commutative where

(a) $\psi : E^{id-1} \times E^{jd-1} \to E^{jd-1} \times E^{id-1}$ is given by $(x,y) \to (y,x)$

and

(b) $E^{jd-1} \times E^{id-1} \xrightarrow{\theta} E^{jd-1} \times E^{id-1}$ is defined as follows:

$$\begin{array}{ccc}
E^{jd-1} \times E^{id-1} & \xrightarrow{\theta} & E^{jd-1} \times E^{id-1} \\
\| & & \| \\
E^{jd-1} \times E^{(i-1)d} \times E^{d-1} & \longrightarrow & E^{jd-1} \times E^{(i-1)d} \times E^{d-1}
\end{array}$$

$$(x,y,z) \longrightarrow (x, h_j(x)^{-1}y, z)$$

$h_j(x)^{-1}y$ is interpreted as follows. View $E^{(i-1)d}$ as $\left\{ \begin{pmatrix} y_1 \\ \vdots \\ y_i \end{pmatrix} \right.$

in $S^{id-1}$: $y_i$ is real and $\geq 0\}$. Then since $j < i$, $h_j(x)^{-1}$ an element of $G(i)$ has the form $\begin{pmatrix} (\ ) & 0 \\ 0 & I \end{pmatrix}$ where $I$ is the $(i-j) \times (i-j)$ identity matrix. In particular

$$h_j(x)^{-1}\begin{pmatrix} y_1 \\ \vdots \\ y_i \end{pmatrix} = \begin{pmatrix} * \\ \vdots \\ * \\ y_i \end{pmatrix}.$$

So if $j < i$, $A \in G(j)$ then the map $E^{(i-1)d} \times E^{(i-1)d}$ defined by $x \to Ax$ is a homeomorphism.

Check commutativity

$$\begin{array}{ccccc} (y,z) \times x & \xrightarrow{\psi} & x \times (y,z) & \xrightarrow{\theta} & x \times (h_j(x)^{-1}y,z) \\ \downarrow h_i \times h_j & & & & \searrow h_j \times h_i \\ h_i(y,z) \times h_j(x) \to h_i(y,z)h_j(x) & \leftarrow & h_j(x) \times h_j(x)^{-1}h_i(y,z)h_j(x) & = & h_j(x) \times h_i(h_j(x)^{-1}y,z) \end{array}$$

$\uparrow$

By Lemma 1.2.4.

We need to calculate $\theta_*$ and $\psi_*$. From Theorem 14 on page 235 of [Spanier] it follows that $\psi_*([i] \times [j]) = (-1)^{(id-1)(jd-1)}([j] \times [i])$

$$= \begin{cases} [j] \times [i] & \text{if } d = 1 \\ -[j] \times [i] & d = 2 \text{ or } 4 \end{cases}.$$

<u>Claim</u>. $\theta_*([j] \times [i]) = [j] \times [i]$. $d = 1,2$ or $4$.

Note that $\theta: E^{jd-1} \times E^{id-1} \to E^{jd-1} \times E^{id-1}$ is a homeomorphism. The inverse is given by $(x,y,z) \to (x,h_j(x)y,z)$. Since $R = Z_2$ for $d = 1$ the claim is true for $d = 1$.

For $d = 2$ or $4$ let $H: E^{jd-1} \times I \to E^{jd-1}$ be a contraction of $E^{jd-1}$ to a point $x_o$ such that $h_j(x_o) = I$. Define $E^{jd-1} \times E^{id-1} \times I \to E^{jd-1} \times E^{id-1}$ by $[(x,y,z),t] \to (x,h_j(H(x,t))^{-1}y,z)$. Thus $\theta$ is homotopic (as a map of pairs) to the map $(x,y,z)$ $\to (x,h_j(x_o)^{-1}y,z) = (x,y,z)$ which proves the claim.

This shows that the stated relations hold in $H_*(G(n):R)$. That these are all the relations can be seen as follows. Let $\Lambda$ be the free associative algebras over $R$ with unit $\underline{1}$ and generated by the elements $i$ of dimension $(id-1)$ where $n \geq i > 0$; subject to the relations

| $d = 1$ | $d = 2$ or $4$ |
|---|---|
| $1 \cdot 1 = \underline{1}$ | |
| $i \cdot i = 0$ if $i > 1$ | $i \cdot i = 0$ for all $i$ |
| $i \cdot j = j \cdot i$ if $i \neq j$ | $i \cdot j = -j \cdot i$ |

Since the corresponding relations hold in $H_*(G(n);R)$, the map $\alpha: \Lambda \to H_*(G(n):R)$ defined by $\underline{1} \to \underline{1}$ and $i \to [i]$ defines a map of algebras. As a module over $R,\Lambda$ is the free $R$ module generated by $\underline{1}$ and the elements $i_1 \cdot i_2 \cdot \ldots \cdot i_r$ such that

$n \geq i_1 > i_2 > \cdots > i_r \geq 0$. Using Lemma 1.3.2, we see that $\alpha$ is an isomorphism of $R$ modules and hence an isomorphism of algebras

The techniques of Section 2 can be generalized to give cell decompositions for the Stiefel Manifolds $G(n,k)$.

Definition. Let $\pi: G(n) \to G(n,k)$ be the projection of a matrix $A$ to the last $(n-k)$ columns of $A$. A normal cell of $G(n,k)$ is by definition the projection of a normal cell $(i_1 \cdots i_r)$ in $G(n)$ to $G(n,k)$ where $n \geq i_1 > i_2 > \cdots > i_r > k$. $\pi(i_1 \cdots i_r)$ is denoted by $(i_1 \cdots i_r | n,k)$.

Using Lemma 1.3.4 and the Lemma's of Section 2, and proceeding by induction on $n$, with $k$ fixed we get

1.3.5(n). $G(n,k)$ is a C.W. complex, whose cells are the normal cells and the 0-cell $\pi(I)$. The map $u\colon Q_n \times G(n-1,k) \to G(n,k)$ $(k < n)$ is cellular and induces an epimorphism of chain complexes.

1.3.6(n). Let $n \geq i_1 \cdots i_r > 0$. Then $\pi(Q_{i_1} \cdots Q_{i_r})$ is contained in the

$$\left(\sum_{s=1}^{r} (i_s d - 1)\right) \text{- skeleton of } G(n,k).$$

1.3.7(n). $u\colon Q_n \times G(n,k) \to G(n,k)$ is cellular.

From 1.3.5(n) and the definition of normal cells we obtain

Corollary 1.3.8. If $m \le n$ and $\ell \le k$, then the induced map $G(m,\ell) \to G(n,k)$ is cellular. This map sends the normal cell $(i_1 \cdots i_r|m,\ell)$ to the normal cell $(i_1 \cdots i_r|n,k)$ if $i_r > k$; for $d = 1$ it sends $(i_1 \cdots i_{r-1}1/m,0)$ to $(i_1 \cdots i_{r-1}/n,k)$ if $i_{r-1} > k \ge 1$. All other cells are mapped degenerately.

Since all differentials in $G(n)$ are zero and $\pi$: $G(n) \to G(n,k)$ induces an epimorphism of chain complexes; all the differentials in $G(n,k)$ are zero.

Definition. If $(i_1 \cdots i_r|n,k)$ is a normal cell, its homology class is denoted by $[i_1 \cdots i_r|n,k]$ and called a normal class. The class of $\pi(I)$ is denoted by $\underline{1}$.

Lemma 1.3.9. $H_*(G(n,k):R)$ is the free $R$ module on the unit class $\underline{1}$ and on the normal classes $[i_1 \cdots i_r|n,k]$. If $m \le n$ and $\ell \le k$, we have a map $G(m,\ell) \to G(n,k)$ which sends $[i_1 \cdots i_r|m,\ell]$ to $[i_1 \cdots i_r|n,k]$ if $i_r > k$; for $d = 1$ it sends $[i_1 \cdots i_{r-1},1|m,0]$ to $[i_1 \cdots i_{r-1}|n,k]$ if $i_{r-1} > k \ge 1$. $\underline{1} \to \underline{1}$ and all other normal classes are mapped to zero. The map $u_*$: $H_*(Q_n \times G(n-1,k):R) \to H_*(G(n,k):R)$ is an epimorphism.

## Exercises

1. Let $G$ be a topological group, $H$ a closed subgroup. Show that if there is a map $f$: $G/H \to G$ such that $\pi \circ f$: $G/H \to G/H$ is the identity on $G/H$, then $G$ is homeomorphic to $G/H \times H$, where $\pi$: $G \to G/H$ is the map $g \to gH$.

2. Let $Q$ denote the quaternions and let $P$ denote the set of pure quaternions. (A quaternion $r = a + cb + jc + kd$ is called pure if $a = 0$, equivalently $r$ is a pure quaternion iff $\bar{r} = -r$.)

Given an element $q$ in $Sp(1)$ define $T_q: Q \to Q$ by $T_q(r) = qrq^{-1}$.

(a) Show $T_q$ maps $P$ onto itself.

Define $f: R^3 \to P$ by $\begin{pmatrix} a \\ b \\ c \end{pmatrix} \to ia + jb + kc$. Given an element $q$ in $Sp(1)$ define the mapping $\theta(q): R^3 \to R^3$ by requiring that

$$\begin{array}{ccc} P & \xrightarrow{T_q} & P \\ f\uparrow & & \downarrow f^{-1} \\ R^3 & \xrightarrow{\theta(q)} & R^3 \end{array}$$

be commutative

(b) Show $\theta(q) \in SO(3)$, and that the map $\theta: Sp(1) \to SO(3)$ is continuous, onto, and a group homomorphism.

(c) Show that the kernel of $\theta$ is the subgroup of $Sp(1)$ consisting of the elements $1$ and $-1$. Conclude that <u>$SO(3)$ is homeomorphic to $RP(3)$</u>.

3. Using 1. and 2. above, show

$O(2)$ is homeomorphic to $S^1 \times S^o$

$O(m)$ is homeomorphic to $SO(m) \times S^o$

$O(4)$ is homeomorphic to $S^3 \times RP^3 \times S^o$

$U(2)$ is homeomorphic to $S^1 \times S^3$

$U(n)$ is homeomorphic to $S^1 \times SU(n)$

4. Prove that if $N$ is a normal subgroup of $O(k+1)$ and also a subgroup of $O(k)$ then $N = \{I\}$, $I$ = identity matrix.

5. Let $X$ be a topological group and let $e$ denote the identity element in $X$. Assume $X$ is a finite C.W. Complex and $e$ is a zero cell. Give $X \times X$ the C. W. Structure determined by the C. W. structure on $X$, and assume that the multiplication $u: X \times X \to X$ is cellular. Let $R$ be a commutative ring with unit and set $C_n(X,R) = H_n(X^n, X^{n-1}:R)$. Prove that $u_*: C_*(X:R) \otimes C_*(X:R) \to C_*(X:R)$ gives $C_*(X:R)$ the structure of an associative algebra with unit $e$.

<u>Notation</u>: Given Topological spaces $X$ and $Y$, let $Y^X$ denote the space of continuous maps of $X$ into $Y$ with the compact open topology. Let $E: Y^X \times X \to Y$ be the evaluation map defined by $E(f,x) = f(x)$.

6. Let $X, Y$, and $Z$ be topological spaces, prove that if $X$ is a locally compact Hausdorff space, then a mapping $g: Z \to Y^X$ is continuous iff the composition $Z \times X \xrightarrow{g\times 1} Y^X \times X \xrightarrow{E} Y$ is continuous.

7. Let $X$ be an arcwise connected topological group with universal covering space $\tilde{X}$. Prove that $\tilde{X}$ is a topological group and

that the projection $\pi\colon \tilde{X} \to X$ is a homomorphism whose kernel is isomorphic to $\pi_1(X,\{e\})$.

8. Show that the fundamental group of an arcwise connected topological group is abelian.

9. If $G$ is a discrete subgroup of the topological group $Y$, then the projection $Y \to Y/G$ is a regular covering space.

10. Describe the following Pontryagin Rings

    (a) $H_*(SO(n):Z_2)$

    (b) $H_*(SO(n):Q)$ $Q$ = rationals

    (c) $H_*(SO(n):Z_p)$ $p$ an odd prime.

11. Let $M(n)$ be the set of $n \times n$ matrices with entries in $F$. ($F$ = reals, complexes, or quaternions) $M(n)$ can be topologized by identifying $M(n)$ with $F^{n^2}$ and giving $F^{n^2}$ the usual topology. On the other hand, the map $M(n) \to (F^n)^{F^n}$ ($(F^n)^{F^n}$ = continuous maps of $F^n$ into itself) defined by $A$ goes to the map $F^n \to F^n$ given by

$$\begin{pmatrix} x_1 \\ \vdots \\ x_n \end{pmatrix} \to A\begin{pmatrix} x_1 \\ \vdots \\ x_n \end{pmatrix} ;$$

is an injection of $M(n)$ into $(F^n)^{F^n}$. Thus $M(n)$ can be topologized as a subspace of $(F^n)^{F^n}$ where $(F^n)^{F^n}$ is given the compact open topology. Show that these two topologies on $M(n)$ are the same. Thus the compact open topology on $GL(n,F)$ and on $G(n)$ is the same as the usual topology on $GL(n,F)$ and on $G(n)$.

12. Show $GL(n,F)$ is an open subset of $F^{n^2}$.

Chapter 2

# FIBRE BUNDLES

## 1. Definition and Examples of Fibre Bundles

*Definition.* Let $F,E,B$ be topological spaces. Let $\pi: E \to B$ be a map. $(F,E,B,\pi)$ is called a *fibre bundle* if there is an open covering $\{U_\alpha\}_{\alpha\in\Lambda}$ of $B$ and for each $\alpha$ in $\Lambda$ a homeomorphism

$\phi_\alpha: U_\alpha \times F \to \pi^{-1}(U_\alpha)$ such that
$$\begin{array}{ccc} \phi_\alpha: U_\alpha \times F & \longrightarrow & \pi^{-1}(U_\alpha) \\ & \underset{(x,u)\to x}{\searrow} & \downarrow \pi \\ & & U_\alpha \end{array}$$
commutes.

$F$ is called the *fibre*, $E$ is called the *total space*, $B$ is called the *base space*. A *map of fibre bundles* is a commutative diagram

$$\begin{array}{ccc} E' & \xrightarrow{f'} & E \\ \downarrow \pi' & & \downarrow \pi \\ B' & \xrightarrow{f} & B \end{array}$$

such that $(\pi')^{-1}\{b'\} \to \pi^{-1}\{f(b')\}$ is a homeomorphism for all $b'$ in $B'$.

The bundle $(F,E',B,\pi')$ is called *equivalent* to $(F,E,B,\pi)$ if there is a map of bundles *with $f'$ a homeomorphism*

$$\begin{array}{ccc} E' & \xrightarrow{f'} & E \\ \downarrow \pi' & & \downarrow \pi \\ B & \xrightarrow{1_B} & B \end{array}$$

Given a fibre bundle $(F,E,B,\pi)$, a map $S: B \to E$ is called a <u>cross section</u> if $\pi \circ S = 1_B$. The bundle $(F, E \times B, B, \pi)$ with $\pi(x,b) = b$ is called the <u>trivial bundle</u>.

The main questions in the theory of fibre bundles are

1. Given a fibre bundle $(F,E,B,\pi)$ are there any cross sections?
2. What are the relations connecting the homology and homotopy groups of the base space, total space, and fibre?
3. Given $B$ and $F$ describe the set of equivalence classes of bundles over $B$ with fibre $F$.

<u>Examples of Fibre Bundles.</u>

1. <u>Möbius Strip</u>

$$I \times I/(0,t) \sim (1,1-t) \longrightarrow I/0 \sim 1 \qquad [t,t'] \longmapsto [t]$$

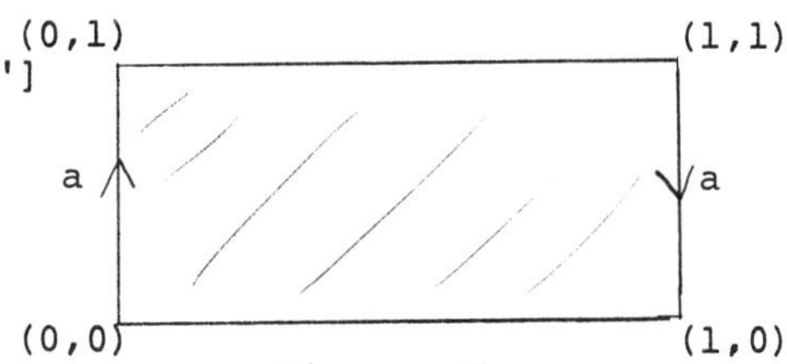


Diagram 2.

The Möbius Strip is not equivalent to the trivial bundle since the Möbius Strip is not homeomorphic to a cylinder

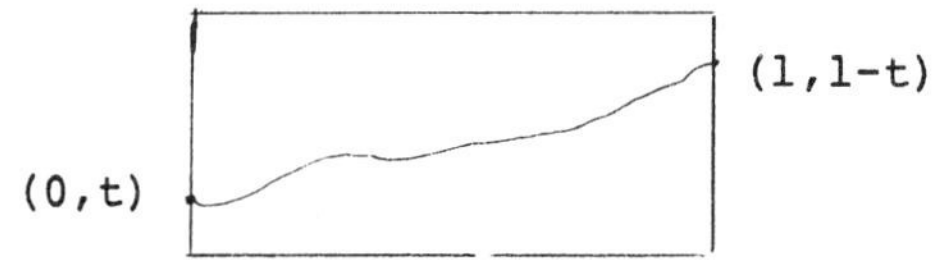

2. <u>Klein Bottle</u>  view $S'$ as $R'/2\pi$

$$\begin{array}{ccc} I \times S'/(0,\theta) \sim (1,2\pi-\theta) & & [t,\theta] \\ \downarrow & & \downarrow \\ I/0 \sim 1 & & [t] \end{array}$$

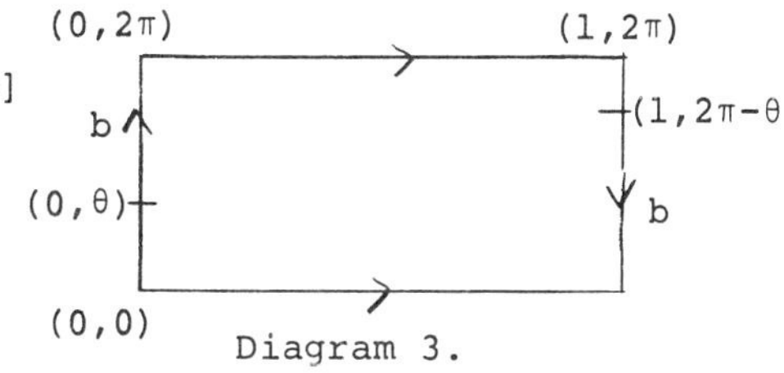

Diagram 3.

This bundle is not equivalent to the trivial bundle since the Klein Bottle is not homeomorphic to a torus

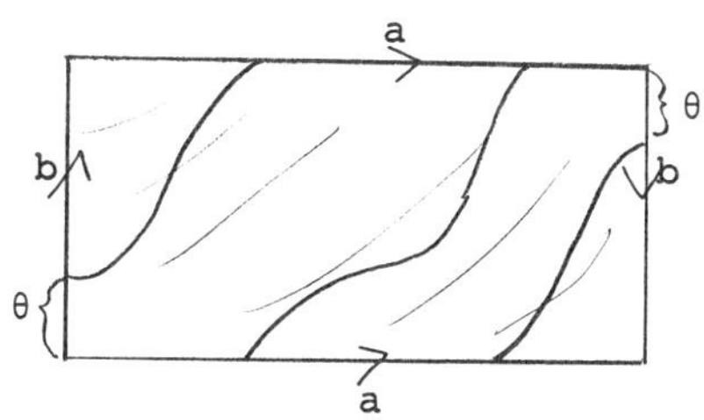


3. <u>Twisted Torus</u>

$$\begin{array}{ccc} I \times S'/(0,\theta) \sim (1,\theta+\pi) & & [t,\theta] \\ \downarrow & & \downarrow \\ I/0 \sim 1 & & [t] \end{array}$$

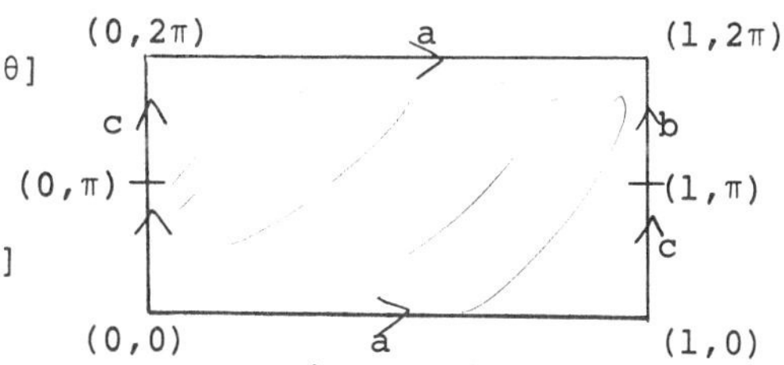

Diagram 4.

The twisted Torus is equivalent to the product bundle. Define $I \times S'/(0,\theta) \sim (1,\theta+\pi) \longrightarrow I \times S'/(0,\theta) \sim (1,\theta)$ by

$$[t,\theta] \to [t,\theta-t\pi].$$

Then

$$[0,\theta] \to [0,\theta]$$

and

$$[1,\theta+\pi] \to [1,\theta+\pi-\pi] = [1,\theta]\ .$$

4. Tangent Bundle

Let $M^n$ be a differentiable manifold then

$$\begin{array}{l} T = \text{space of vectors tangent to } M^n \\ \Big\downarrow \pi \\ M^n \text{ is a bundle with fibre } R^n. \end{array}$$

5. Normal Bundle

Let $M^n$ be differentiably embedded in $R^{n+k}$, then

$$\begin{array}{l} \nu = \text{normal vectors to } M^n \text{ in } R^{n+k} \\ \Big\downarrow \pi \\ M^n \text{ is a bundle with fibre } R^k. \end{array}$$

6. Projection Spaces. Let $d = 1,2,$ or $4.$

Define an equivalence relation on $S^{(n+1)d-1}$ by

$$\begin{pmatrix} x_1 \\ \vdots \\ x_{n+1} \end{pmatrix} \sim \begin{pmatrix} x_1\lambda_1 \\ \vdots \\ x_{n+1}\lambda \end{pmatrix} .$$

for $\lambda \in S^{d-1}$. Let $FP^n$ denote the quotient space.

Then $S^{d-1} \longrightarrow S^{(n+1)d-1}$
$$\downarrow$$
$FP^n$ = "lines" in $F^{n+1}$ is a fibre bundle.

To see that this is a fibre bundle, set $U_i = \left\{ \begin{pmatrix} x_1 \\ \vdots \\ x_{n+1} \end{pmatrix} \text{ in } FP^n : x_i \neq 0 \right\}$, $U_i$ is open in $FP^n$ and $U_i$ $i=1\cdots n+1$ cover $FP^n$.

Define $U_i \times S^{d-1} \to S^{(n+1)d-1}$ by $\begin{pmatrix} x_1 \\ \vdots \\ x_{n+1} \end{pmatrix} \times \lambda \to \begin{pmatrix} y_1\lambda \\ \vdots \\ y_{n+1}\lambda \end{pmatrix}$ where

$\begin{pmatrix} y_1 \\ \vdots \\ y_{n+1} \end{pmatrix}$ is the unique vector in $S^{(n+1)d-1}$ which projects to

$\begin{pmatrix} x_1 \\ \vdots \\ x_{n+1} \end{pmatrix}$ and has $y_i$ real and $> 0$. $FP^n$ is an nd-dimensional, closed, differentiable manifold.

If $d = 1$ $FP^n = RP^n$ = n dimensional Real Projective Space

$d = 2$ $FP^n = CP^n$ = n dimensional Complex Projective Space

$d = 4$ $FP^n = QP^n$ = n dimensional Quaternionic Projective Space.

## 7. Homogeneous Spaces

Let $H_o$ be a closed subgroup of $H_1$, and let $H_1$ be a

closed subgroup of G. Under mild restrictions $\begin{array}{c} G/H_0 \\ \downarrow \\ G/H_1 \end{array}$ is a fibre bundle with fibre $H_1/H_0$. In particular, the following are fibre bundles

$$\begin{array}{ccc} G(k) & \longrightarrow & G(n) \\ & & \downarrow \\ & & G(n,k) \end{array} \qquad \begin{array}{ccc} G(n-k) & \longrightarrow & G(n,k) \\ & & \downarrow \\ & & M(n,k) \end{array}$$

and

$$\begin{array}{ccc} G(k+i,k) & \longrightarrow & G(n,k) \\ & & \downarrow \\ & & G(n,k+i) \end{array}$$

8. <u>Covering Spaces (For details see [Massey])</u>

Assume all spaces are arcwise connected and locally arcwise connected.

<u>Definition</u>. $\begin{array}{c} \tilde{X} \\ \downarrow P \\ X \end{array}$ is a covering space if for each point x in X there is an arcwise connected open nbhd U of x such that the projection P restricted to any arc component of $p^{-1}(U)$ is a homeomorphism onto U.

The sets $P^{-1}(x)$ for each $x \varepsilon X$ and the coset space $\pi_1(X,x)/P_*\pi_1(\tilde{X},\tilde{x})$ have the same cardinality. So a covering space is an example of a fibre bundle with discrete fibre.

1. Lifting Problem (cross sections).

$$\begin{array}{ccc} & & (\tilde{X},\tilde{x}) \\ & \overset{\tilde{\phi}}{\nearrow} & \downarrow \\ (Y,y) & \xrightarrow{\phi} & (X,x) \end{array}$$

a lifting $\tilde{\phi}$ exists iff $\phi_*\pi_1(Y,y) \subset P_*\pi(\tilde{X},\tilde{x})$. Moreover, if a lifting $\tilde{\phi}$ exists then it is unique.

It follows that $0 \to \pi_1(\tilde{X},\tilde{x}) \xrightarrow{P_*} \pi_1(X,x)$ is exact and hence the only covering space of $X$ which admits a cross section is the covering space $\begin{array}{c} X \\ \downarrow 1_X \\ X \end{array}$.

2. Recall that for $n \geq 0$, $\pi_n(X,x)$ is the set $[(S^n,u_1),(X,x)]$ = base point preserving homotopy classes of maps. For $n \geq 1$, $\pi_n(X,x)$ can be given a group structure and for $n \geq 2$, $\pi_n(X,x)$ is an abelian group.

Since, cardinality of $p^{-1}(X)$ = cardinality $\pi_1(X,x)/P_*\pi(\tilde{X},\tilde{x})$ we have

$$0 \to \pi_1(\tilde{X},\tilde{x}) \to \pi_1(X,x) \to \pi_o(F) = \pi(X,x)/P_*\pi(\tilde{X},\tilde{x}) \to 0 \text{ is exact.}$$

Since $\pi_1(S^n,u_1) = 0$ for $n \geq 2$, we have from the lifting problem that $\pi_n(\tilde{X},\tilde{x}) \xrightarrow[\approx]{P_*} \pi_n(X,x)$ for $n \geq 2$.

3. Two covering spaces $\tilde{X}_o$ and $\tilde{X}_1$ are called equivalent if there is a homeomorphism $h\colon \tilde{X}_o \to \tilde{X}_1$ such that

$$\begin{array}{ccc} \tilde{X}_o & \longrightarrow & \tilde{X}_1 \\ & {\scriptstyle P_o}\searrow \quad \swarrow{\scriptstyle P_1} & \\ & X & \end{array}$$

commutes. The equivalence classes of coverings of $X$ are in a 1-1 correspondence with the conjugacy classes of subgroups of $\pi_1(X,x)$. [Providing $X$ has a universal covering space.]

## 2. Pullbacks

Lemma 2.2.1. Let $\begin{array}{ccc} E' & \xrightarrow{f} & E \\ \downarrow{\scriptstyle \pi'} & & \downarrow{\scriptstyle \pi} \\ B & \xrightarrow{1_B} & B \end{array}$ be a map of bundles. If the fibre $F$ is a locally connected, locally compact Hausdorff space, then $f$ is an equivalence of fibre bundles.

Proof. $f$ is 1-1 and onto since it is a map of bundles. So the proof reduces to showing that $f^{-1}$ is continuous. This can be reduced to the following "local" problem. Let $\begin{array}{ccc} U \times F & \xrightarrow{f} & U \times F \\ & {\scriptstyle P_1}\searrow \quad \swarrow{\scriptstyle P_1} & \\ & U & \end{array}$

be commutative. Define $f_u: F \to F$ for each $u \in U$ by $f_u(x) = P_2 \circ f(u,x)$. Prove that if each $f_u$ is a homeomorphism, and if $F$ is a locally connected, locally compact Hausdorff space; then $f^{-1}$ is continuous.

Give $F^F$ the compact open topology (see Appendix 2), then

1. If $F$ is a locally compact Hausdorff space, then a map $f: U \times F \to F$ is continuous iff the map $U \to F^F$ defined by $u \to f_u$ is continuous.

2. Let $h(F^F)$ denote the subspace of $F^F$ consisting of homeomorphisms of $F$ onto $F$. If $F$ is a locally connected, locally compact Hausdorff space, then $h(F^F)$ is a topological group.

The rest of the proof proceeds as follows

The map $U \times F \to F$ defined by $(u,x) \to f_u(x)$ is continuous so by 1. above, the map $U \to F^F$ defined by $u \to f_u$ is continuous. But each $f_u$ is a homeomorphism, so $U \to h(F^F)$ defined by $u \to f_u$ is continuous. By 2. above, $h(F^F)$ is a topological group and so the composition $U \to h(F^F) \xrightarrow{\text{take inverse}} h(F^F)$, $(u \to f_u \to f_u^{-1})$ is continuous. Using 1. again we have that $U \times F \to F$ defined by $(u,x) \to f_u^{-1}(x)$ is continuous. So the map $U \times F \to U \times F$ given by $(u,x) \to (u,f_u^{-1}(x))$ is continuous. But this last map is $f^{-1}$ and so we are done.

<u>Definition</u>. Let $\begin{array}{c} E \\ \downarrow{\scriptstyle \pi} \\ B \end{array}$ be a fibre bundle, and $f\colon B' \to B$ a map. Set $E_f$ equal to the subspace of $B' \times E$ consisting of points $(b',x)$ such that $f(b') = \pi(x)$. $E_f$ is called the pullback of $E$ over $f$. Define $E_f \xrightarrow{\pi'} B'$ by $(b',x) \to b'$ and $E_f \xrightarrow{\pi'} E$ by $(b',x) \to x$, then

$$\begin{array}{ccc} E_f & \xrightarrow{f'} & E \\ \downarrow{\scriptstyle \pi'} & & \downarrow{\scriptstyle \pi} \\ B' & \xrightarrow{f} & B \end{array}$$

commutes.

<u>Lemma 2.2.2</u>. The pullback of a fibre bundle is a fibre bundle and

$$\begin{array}{ccc} E_f & \xrightarrow{f'} & E \\ \downarrow{\scriptstyle \pi'} & & \downarrow{\scriptstyle \pi} \\ B' & \xrightarrow{f} & B \end{array}$$

is a map of bundles.

<u>Proof</u>. This follows since the pullback of the trivial bundle is equivalent to the trivial bundle. If $\{U_j,\phi_j\}$ is a collection of coordinate nbhds and functions, then $\{f^{-1}(U_j),\overline{\phi}_j\}$ is a collection of coordinate nbhds and functions where $\overline{\phi}_j : f^{-1}U_j \times F \to E'$ is

defined by $\overline{\phi}_j(u,x) = (u,\phi_j(f(u),x))$.

Lemma 2.2.3. If

$$\begin{array}{ccc} E' & \xrightarrow{f'} & E \\ \downarrow & & \downarrow \\ B' & \xrightarrow{f} & B \end{array}$$

is a map of bundles with fibre $F$, a locally connected, locally compact Hausdorff space, then $E'$ is equivalent to the pullback $E_f$.

Proof. Use Lemma 2.2.1.

Examples of pullbacks

Let $V^n$ and $W^n$ be differentiable manifolds, $f\colon V^n \to W^n$ a differentiable map. $f$ induces a map $f'\colon T(V) \to T(W)$ with

$$\begin{array}{ccc} T(V) & \xrightarrow{f'} & T(W) \\ \downarrow & & \downarrow \\ V & \xrightarrow{f} & W \end{array}$$

commutative. But $f'$ is not in general a map of fibre bundles, because it may fail to be a homeomorphism on some fibres.

Write $f(x) = f(x_1 \cdots x_n) = (f_1(x),\ldots f_n(x))$ for $x$ near $v_o$ (i.e. in $U_{v_o}$). Then the rank of the Jacobian matrix $\left(\frac{\partial f_i}{\partial x_j}\right)_{i,j}$ at

$v_o$ is independent of the choice of coordinate patches.

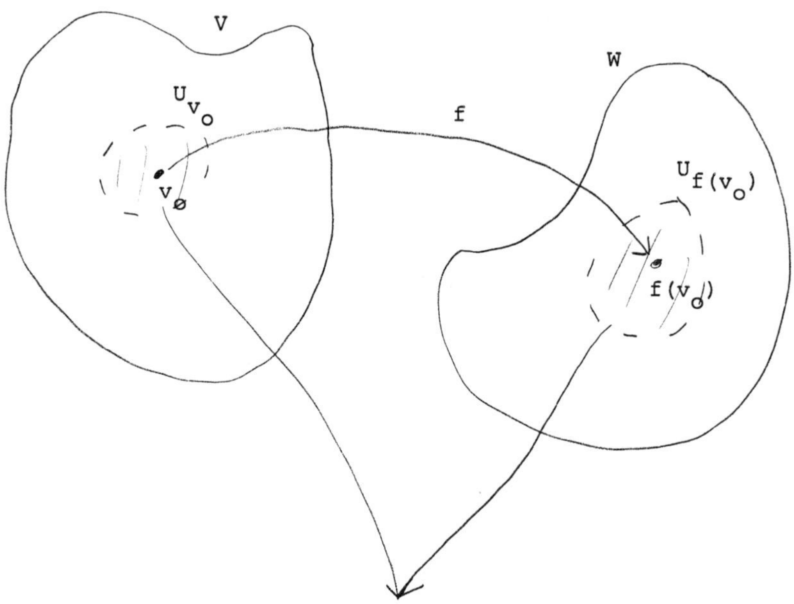


Coordinate Patches

(Diagram 5)

It follows that <u>f' is a map of fibre bundles iff</u> <u>det</u> $\left|\frac{\partial f_i}{\partial x_j}\right|_{v_o} \neq 0$. Let $M^n$ be a differentiable Manifold and $M^n \to R^{n+k}$ an embedding. Recall $M(n+k,k)$ is the space of n planes in $R^{n+k}$. Set $E(n+k,k)$ equal to the space of pairs (n plane in $R^{n+k}$, vector in that plane). Then

$$\begin{array}{ccc} T(M) & \longrightarrow & E(n+k,k) \\ \downarrow & & \downarrow \\ M^n & \longrightarrow & M(n+k,k) \end{array}$$

is a map of bundles with fibre $R^n$. $M^n \to M(n+k,k)$ defined by $x \to$ tangent plane at $x$ and

$$\begin{array}{ccc} v & \longrightarrow & E(n+k,n) \\ \downarrow & & \downarrow \\ M^n & \longrightarrow & M(n+k,n) \end{array}$$

is a map of bundles with fibre $R^k$. $v$ = normal bundle of the immersion $M^n \to M(n+k,n)$ defined by $x \to$ normal plane at $x$.

Definition. Let $\begin{array}{c} E \\ \downarrow \pi \\ B \end{array}$ and $\begin{array}{c} E' \\ \downarrow \pi' \\ B' \end{array}$ be bundles with fibres $F$ and $F'$.

Then $\begin{array}{c} E \times E' \\ \downarrow \pi\times\pi' \\ B \times B' \end{array}$ is a fibre bundle with fibre $F \times F'$. $E \times E'$ is called the product bundle.

Definition. Let $E$ and $E'$ be fibre bundles over $B$ with fibres $F, F'$.

The Whitney Sum $E \oplus E'$ is defined to be the pullback

$$\begin{array}{ccc} E \oplus E' & \longrightarrow & E \times E' \\ \downarrow & & \downarrow \\ B & \xrightarrow[\Delta]{} & B \times B \end{array} \qquad \text{where} \quad \Delta(b) = (b,b)$$

If $M^n \to R^{n+k}$ is an embedding, then $T \oplus \nu$ is equivalent to the trivial bundle

$$\begin{array}{ccccc} T \oplus \nu & \longrightarrow & E(n+k,0) & = & R^{n+k} \\ \downarrow & & \downarrow & & \downarrow \\ M^n & \longrightarrow & M(n+k,0) & = & pt \end{array}$$

## 3. Classification of Bundles in Terms of Coordinate Transformations

Throughout this section, we will assume $F$ is a locally connected, locally compact, Hausdorff space.

So that

(a) homeomorphisms $f: U \times F \to U \times F$ with

$$\begin{array}{ccc} U \times F & \longrightarrow & U \times F \\ & {\scriptstyle p_1} \searrow \quad \swarrow {\scriptstyle p_1} & \\ & U & \end{array}$$

commutative are in a 1-1 correspondence with continuous maps $f: U \to h(F)$ where $h(F)$ denotes the set of homeomorphisms of $F$ onto itself, given the compact open topology.

(b) $h(F)$ is a topological group.

Let $\begin{array}{c} E \\ \downarrow \pi \\ B \end{array}$ be a fibre bundle with fibre $F$. Then we can choose an open covering $\{U_j\}_J$ of $B$ and for each $j$ in $J$ a homeomorphism $\phi_j\colon U_j \times F \to \pi^{-1}(U_j)$ with

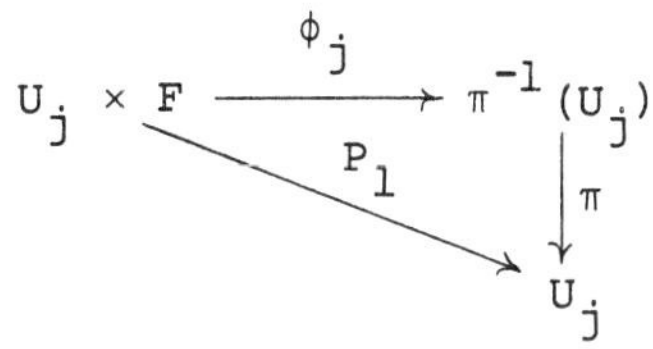


commutative. If $U_i \cap U_j \neq \emptyset$, then we have

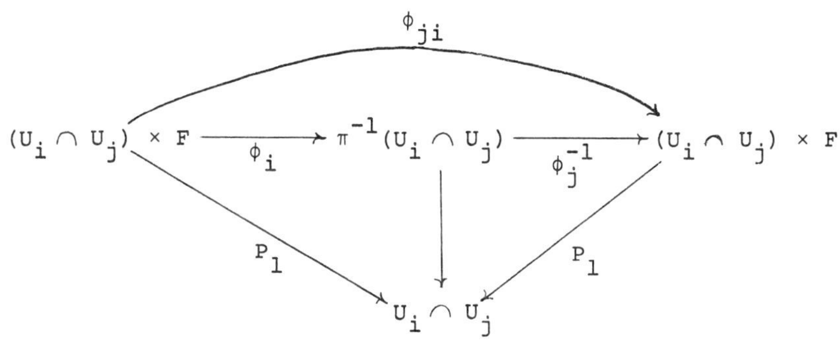


Since the diagram commutes, we can view $\phi_{ji}$ as a map $\phi_{ji}\colon (U_i \cap U_j) \times F \to F$, or $\phi_{ji}(U_i \cap U_j) \to h(F)$.

Given the open covering $\{U_j\}$ of $B$ and the maps $\phi_{ji}\colon U_j \cap U_i \to h(F)$. We can construct a bundle $E'$, equivalent to $E$ from the disjoint union $\bigcup_{j \in J} (U_j \times F)$ by setting $(u,x) \sim (u,\phi_{ji}(u)\cdot x)$, $u \in U_j \cap U_i$ and setting $E'$ equal to the corresponding quotient space (a proof follows the examples).

Examples.

(1) Define $U_1 \subset S'$ and $U_2 \subset S'$ by

$$U_1 = \{\theta: -\frac{1}{10} < \theta < \pi + \frac{1}{10}\}$$

$$U_2 = \{\theta: -\pi - \frac{1}{10} < \theta < \frac{1}{10}\}$$

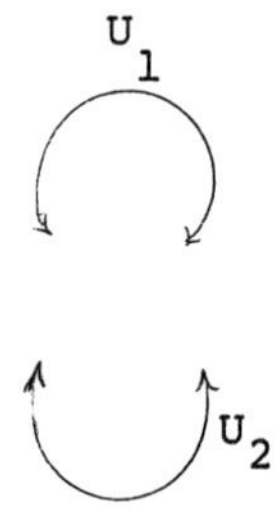


Then $U_1 \cap U_2 = (\pi - \frac{1}{10}, \pi + \frac{1}{10}) \cup (-\frac{1}{10}, \frac{1}{10})$.

Define

$$U_1 \cap U_2 \to h([0,1])$$

$$(-\frac{1}{10}, \frac{1}{10}) \to e = \text{identity}$$

$$(\pi - \frac{1}{10}, \pi + \frac{1}{10}) \to f \quad \text{where} \quad f(t) = 1 - t.$$

This gives the Möbius Strip.

(2) Same $U_1$ and $U_2$ as in (1)

Define

$$U_1 \cap U_2 \to h(S')$$

$$(-\frac{1}{10}, \frac{1}{10}) \to e$$

$$(\pi - \frac{1}{10}, \pi + \frac{1}{10}) \to g \quad \text{where} \quad g(\theta) = -\theta.$$

This gives the Kline Bottle.

(3) Same $U_1$ and $U_2$ as in (1)

Define

$$U_1 \cap U_2 \to h(S')$$

$$(-\frac{1}{10}, \frac{1}{10}) \to e$$

$$(\pi - \frac{1}{10}, \pi + \frac{1}{10}) \to g' \quad \text{where} \quad g'(\theta) = \pi + \theta.$$

This gives the twisted Torus.

(4) Let $M^n$ be a differentiable Manifold.

Let $\{U_j\}_{j \varepsilon J}$ be an open cover of $M^n$ by coordinate patches.

Define $g_{ji}: (U_i \cap U_j) \to h(R^n)$ as follows.

$f$ defined on points corresponding to $U_i \cap U_j$.

Let $u$ be a point in $U_i \cap U_j$ and set $g_{ji}(u)$ equal to the linear transformation $R^n \to R^n$ whose matrix with respect to the

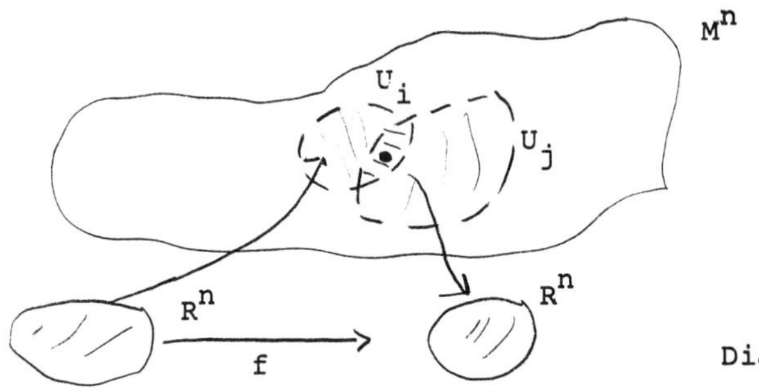


Diagram 6.

standard basis for $R^n$ is $\left(\frac{\partial f_i}{\partial x_j}\right)_{i,j}$ evaluated at $u$. This gives the tangent bundle to $R^n$.

Suppose we are given an open covering $\{U_j\}_J$ of $B$ and a collection of continuous maps $\phi_{ji}: U_j \cap U_i \to h(F)$. Give $J$ the discrete topology and set $T$ equal to the subspace of $B \times F \times J$ consisting of triples $(u,x,j)$ such that $u \in U_j$ ($T$ is just the disjoint union $\bigcup_j (U_j \times F)$). Define an equivalence relation on $T$ by requiring $(u,x,j) \sim (u,\phi_{kj}(u)\cdot x,k)$ for $u \in U_k \cap U_j$. Set $E_{\{U_j,\phi_{ji}\}}$ equal to the corresponding quotient space, and let $q: T \to E_{\{U_j,\phi_{ji}\}}$ be the projection $(u,x,j) \to [u,x,j]$. Define $\tilde{\pi}: E \to B$ by $\tilde{\pi}[u,x,j] = u$, then $\tilde{\pi}$ is well defined and continuous.

For each $j \in J$, define $q_j: U_j \times F \to \tilde{\pi}^{-1}(U_j)$ by $(u,x) \to [u,x,j]$. Then $q_j$ is continuous and

$$\begin{array}{ccc} q_j\colon U_j \times F & \longrightarrow & \tilde{\pi}^{-1}(U_j) \\ & \searrow^{P_1} & \downarrow \tilde{\pi} \\ & & U_j \end{array}$$

commutes. Moreover, $q_j$ is onto $([u,x,k]$ with $u \in U_j \cap U_k$ is a typical element of $\pi^{-1}(U_j)$, $[u,x,k] = [u,\phi_{jk}(u)\cdot x,j] = q_j(u,\phi_{jk}(u)\cdot x))$.

Find a sufficient condition on the $\phi_{ji}$ which will guarantee that each $q_j\colon U_j \times F \to \tilde{\pi}^{-1}(U_j)$ is a homeomorphism.

Claim (*) $g_{kj}(u)g_{ji}(u) = g_{ki}(u)$, $u \in U_k \cap U_j \cap U_i$ implies that each $q_j$ is a homeomorphism.

Proof. Observe that (*) implies $g_{ii}(u) = e$ identity

$$g_{ij}(u) = [g_{ji}(u)]^{-1}$$

Note also that $\sim$ is now an equivalence relation.

(a) $(u,x,j) \sim (u,g_{kj}(u)\cdot x,k) \sim (u,g_{\ell k}(u)g_{kj}(u)\cdot x,\ell)$
$= (u,g_{\ell j}(u)\cdot x,\ell)$ so $x \sim y$ and $y \sim z \Rightarrow x \sim z$.

(b) $(u,x,j) \sim (u,g_{kj}(u)\cdot x,k) \sim (u,g_{jk}(u)g_{kj}(u)\cdot x,j) = (u,x,j)$
so $x \sim y \Rightarrow y \sim x$

(c) $(u,x,j) = (u,g_{jj}(u)x,j)$ so $x \sim x$.

$g_j\colon U_j \times F \to \tilde{\pi}^{-1}(U_j)$ is 1-1.

If $[u,x,j] = [u,x',j]$ then

$$x' = g_{ji_1}(u) \cdot g_{i_1 i_2}(u) \cdot \cdots \cdot g_{i_{n-1} i_n}(u) g_{i_n j}(u) \cdot x$$

$$= g_{jj}(x) = x$$

so $q_j$ is 1-1.

It remains to show that if $W$ is an open subset of $U_j \times F$ then $g_j(W)$ is open in $E_{\{U_j, \phi_{ji}\}}$, which is equivalent to showing that $q^{-1}q_j(W)$ is open in $T$. Since $\bigcup_j U_j \times F \times \{j\}$ is an open cover of $T$, it suffices to show that $g^{-1}q_j(W) \cap (U_k \times F \times \{k\})$ is open for all $k$

$$\begin{array}{ccc} (U_j \cap U_k) \times F \times \{k\} & \xrightarrow{\text{inclusion}} & T \\ \Big\downarrow \tilde{S} & & \Big\downarrow q \\ U_j \times F & \xrightarrow{\quad q_j \quad} & E \end{array}$$

commutes where $\tilde{S}(u,x,k) = (u, g_{jk}(u) \cdot x)$. Using our assumptions on $F$, we see that $\tilde{S}$ is continuous so $\tilde{S}^{-1}(W)$ is open in $(U_j \cap U_k) \times F \times \{k\}$. But since the diagram commutes we have $\tilde{S}^{-1}(W) = q^{-1}q_j(W) \cap ((U_j \cap U_k) \times F \times k)$ and we are done. Note $q^{-1}q_j(W) \cap (U_k \times F \times \{k\}) = q^{-1}q_j(W) \cap ((U_j \cap U_k) \times F \times \{k\})$.

<u>Definition</u>. By a system of coordinate transformations in $B$ with values in $h(F)$ is meant an indexed covering $\{U_j\}_J$ of $B$ by

open sets and a collection of continuous maps $\phi_{ji}\colon U_j \cap U_i \to h(F)$ such that $g_{kj}(u)\cdot g_{ji}(u) = g_{ki}(u)$ if $u \in U_i \cap U_j \cap U_k$.

<u>Definition</u>. $(F,E,B,\pi,\{U_j\},\{\phi_j\})$ is called a coordinate bundle if $(F,E,B,\pi)$ is a fibre bundle, $\{U_j\}$ is an indexed open cover of $B$ and for each $j \in J$, $\phi_j\colon U_j \times F \to \pi^{-1}(U_j)$ is a homeomorphism with

$$\begin{array}{ccc} \phi_j\colon U_j \times F & \longrightarrow & \pi^{-1}(U_j) \\ & {\scriptstyle p_1}\searrow & \downarrow{\scriptstyle \pi} \\ & & U_j \end{array}$$

commutative.

The discussion above proves.

<u>Lemma 2.3.1</u>. If $\{U_j\},\{\{\phi_{ji}\}$ is a system of coordinate transformations in $B$ with values in $h(F)$, then $(F,E_{\{U_j,\phi_{ji}\}},B,\tilde{\pi},\{U_j\},\{q_j\})$ is a coordinate bundle.

<u>Lemma 2.3.2</u>. Let $(F,E,B,\pi,\{U_j\},\{\phi_j\})$ be a coordinate bundle. Define

$$\{\phi_{ij}\}_{(j,i)} \quad \text{in} \quad J \times J \quad \text{with} \quad U_j \cap U_i \neq \emptyset$$

by

$$(U_i \cap U_j) \times F \xrightarrow{\ \phi_i\ } \pi^{-1}(U_i \cap U_j) \xrightarrow{\ \phi_j^{-1}\ } (U_i \cap U_j) \times F$$

$$\phi_{ji} \text{ (the composite, from } (U_i \cap U_j) \times F \text{ to } (U_i \cap U_j) \times F)$$

then $\{U_j\}$, $\{\phi_{ji}\}$ is a system of coordinate transformations in $B$ with values in $h(F)$.

Proof.

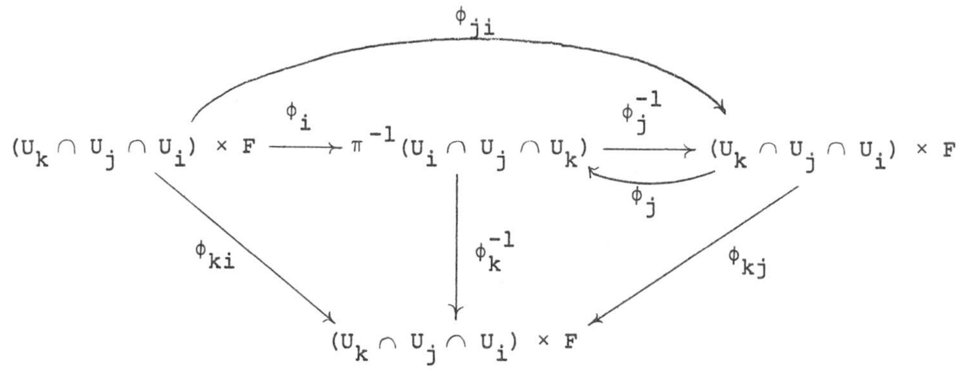


commutes so $\phi_{kj}\phi_{ji} = \phi_{ki}$ if $(U_i \cap U_j \cap U_k) \neq \emptyset$.

Lemma 2.3.3. Let $(F,E,B,\pi, \{U_j\},\{\phi_j\})$ be a coordinate bundle, if $\{U_j,\phi_{ji}\}$ denotes the corresponding system of coordinate transformations then $(F,E,B,\pi)$ and $(F,E_{\{U_j,\phi_{ji}\}},B)$ are equivalent fibre bundles.

Proof. Let $T$ be the subspace of $B \times F \times J$ consisting of triples $(u,x,j)$ such that $u \in U_j$. Define $\phi\colon T \to E$ by $\phi(u,x,j) = \phi_j(u,x)$. To show that if $(u,x,j) \sim (u,x',k)$ then $\phi(u,x,j) = \phi(u,x',k)$ it suffices to show that $\phi(u,x,j) = \phi(u,\phi_{kj}(u)\cdot x,k)$ if $u \in U_k \cap U_j$.

But $\phi_j(u,x) = \phi_k(u,\phi_{kj}(u)\cdot x)$ follows from the identity $\phi_k^{-1}\phi_j(u,x) = (u,\phi_{kj}(u)\cdot x)$.

Thus $\phi$ induces a map $\phi_*: E \xrightarrow[\{U_j,\phi_{ji}\}]{} E$ with

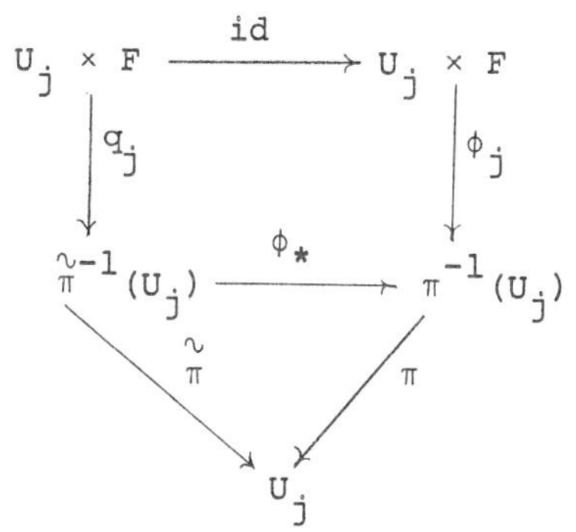


commutative. So $\phi_*$ is an equivalence of bundles.

Using Lemmas 2.3.1, 2.3.2, 2.3.3 we get a commutative diagram

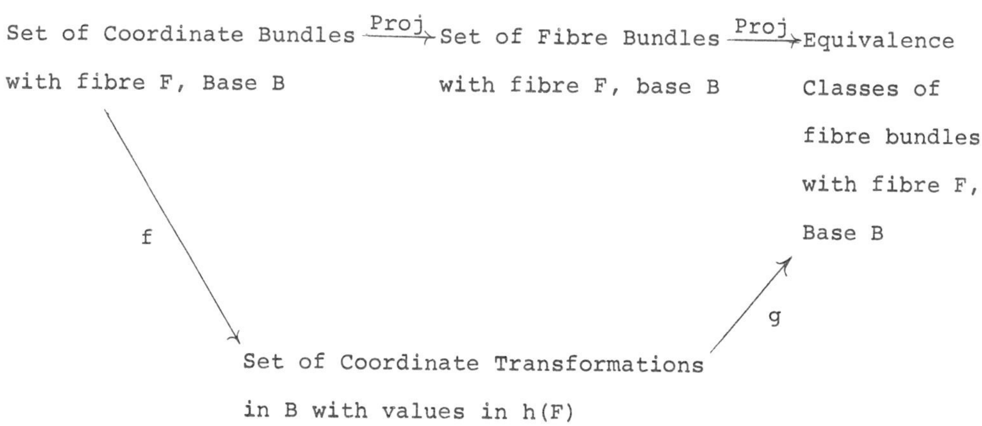

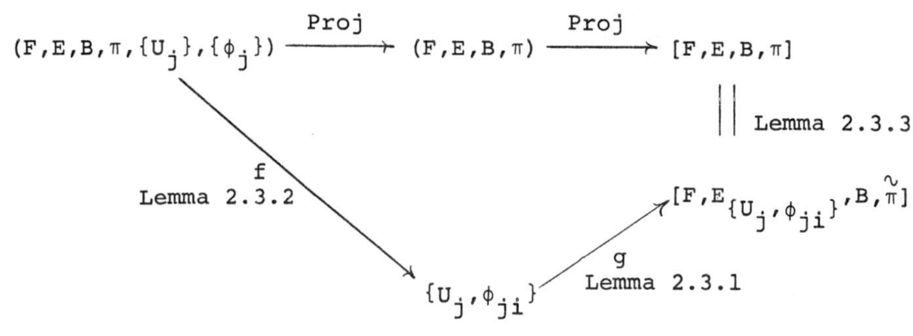


Since $(\mathrm{Proj})\circ(\mathrm{Proj})$ is onto, it follows that $g$ is onto. Thus one can define on the set of coordinate transformations an equivalence relation so that the equivalence classes of coordinate transformations are in a 1-1 correspondence with the equivalence classes of fibre bundles.

We proceed to define this equivalence relation. Suppose $\{U_j,\phi_{ji}\}$ and $\{U_{j'},\phi_{j'i'}\}$ are systems of coordinate transformations with corresponding coordinate bundles $(F,E,B,\pi,\{U_j\},\{q_j\})$ and $(F,E',B,\pi',\{U_{j'}\},\{q_{j'}\})$. Let $f\colon E' \to E$ be an equivalence of bundles. It is not hard to see that $\{U_{j'}\}$, $\{f\circ q_{j'}\}$ is a system of coordinates nbhds, and coordinate functions for the bundle $E$. So $(\{U_j\} \cup \{U_{j'}\})_{J \cup J'}$, $(\{q_j\} \cup \{q_{j'}\})_{J \cup J'}$ is a system of coordinate nbhds and coordinate functions for the bundle $E$. Denote the corresponding system of coordinate transformations by $(\{U_j\} \cup \{U_{j'}\}, \bar{\phi}_{ji}, \bar{\phi}_{ji'}, \bar{\phi}_{j'i}, \bar{\phi}_{j'i'})$. It follows that

$$\phi_{ji} = \bar{\phi}_{ji}\colon U_i \cap U_j \to h(F)$$

$$\phi_{j'i'} = \bar{\phi}_{j'i'}\colon U_{i'} \cap U_{j'} \to h(F).$$

The new information is given by the maps $\overline{\phi}_{j',i}\colon U_{j'} \cap U_i \to h(F)$ which must satisfy

eq. 1. $\overline{\phi}_{k'i} = \phi_{k'j'} \circ \overline{\phi}_{j'i}$ on $U_{k'} \cap U_{j'} \cap U_i$

and

eq. 2. $\overline{\phi}_{k'i} = \overline{\phi}_{k'j} \circ \phi_{ji}$ on $U_{k'} \cap U_j \cap U_i$

This motivates

<u>Definition</u>. Let $\{U_{j'}, \phi_{ji}\} \sim \{U_{j'}, \phi_{j'i'}\}$ be systems of coordinate transformations in B with values in h(F). Set $\{U_j, \phi_{ji}\}$ $\sim \{U_{j'}, \phi_{j'i'}\}$ iff there exist maps $\phi_{j'i}\colon U_{j'} \cap U_i \to h(F)$ satisfying

eq. 1'. $\phi_{k'i} = \phi_{k'j'} , \phi_{j'i}$ on $U_{k'} \cap U_{j'} \cap U_i$

and

eq. 2'. $\phi_{k'i} = \phi_{k'j} \circ \phi_{ji}$ on $U_{k'} \cap U_j \cap U_i$

<u>Remarks</u> 1. If $\{U_j, \phi_{ji}\} \sim \{U_{j'}, \phi_{j'i'}\}$ then $(\{U_j\} \cup \{U_{j'}\},$ $\phi_{ji}, \phi_{j',i'}, \phi_{j'i}, \phi_{ij'})$ is a system of coordinate transformations where $\phi_{ij'}$ is defined to be $\phi^{-1}_{j',i}$. So $\{U_{j'}, \phi_{ji}\} \sim \{U_{j'}, \phi_{j',i'}\}$ iff $\phi_{j',i}$ and $\phi_{i,j'}$ can be defined so that $(\{U_j\} \cup \{U_{j'}\}, \phi_{ji}, \phi_{j',i'}, \phi_{j',i}, \phi_{ij'})$ forms a system of cooordinate transformations.

2. $\sim$ is an equivalence relation

$$\{U_j,\phi_{ji}\} \sim \{U_j,\phi_{ji}\} \quad \text{O.K.}$$

$$\{U_j,\phi_{ji}\} \sim \{U_{j'},\phi_{j',i'}\} \Longrightarrow \{U_{j'},\phi_{j',i'}\}$$

$$\sim \{U_j,\phi_{ji}\} \quad \text{clear from definition.}$$

To see that $\{U_j,\phi_{ji}\} \sim \{U_{j'},\phi_{j',i'}\} \sim \{U_{j''},\phi_{j'',i''}\} \Longrightarrow \{U_j,\phi_{ji}\}$ $\sim \{U_{j''},\phi_{j''i''}\}$, define $\phi_{j'',i}$ as follows.

Let $u$: $U_{j''} \cap U_i$, then $u \in U_{k'}$ for some $k'$ set $\phi_{j''i}(u) = \phi_{j'',k'}(u)\phi_{k'i}(u)$. This is well defined since if $u \in U_{k'} \cap U_{\ell'}$ we have

$$\phi_{j''k'}(u)\phi_{k',i}(u) = \phi_{j''k'}(u)\phi_{k',\ell'}(u)\phi_{\ell'k'}(u)\phi_{k'i}(u)$$

$$= \phi_{j''\ell'}(u)\phi_{\ell',i}(u)$$

equation 1', and equation 2' follow directly.

3. If $\{U_j,\phi_{ji}\}$ and $\{U_{j'},\phi_{j',i'}\}$ determine equivalent fibre bundles then $\{U_{j'},\phi_{ji'}\} \sim \{U_j,\phi_{ji}\}$.

To show that the converse of 3. is also true it suffices to show that if $\{U_j,\phi_{ji}\} \sim \{U_{j'},\phi_{j',i'}\}$ then $\{U_j,\phi_{ji}\}$ and $(\{U_j\} \cup \{U_{j'}\},$ $\phi_{ji},\phi_{j',i'},\phi_{j',i})$ determine equivalent fibre bundles.

Let $(F,E,B,\pi,\{U_j\},\{q_j\})$ and $(F,\overline{E},B,\overline{\pi},\{U_j\} \cup \{U_{j'}\},\{\overline{q}_j\} \cup \{\overline{q}_{j'}\})$ denote the corresponding coordinate bundles. $[u,x,j] \to [u,x,j]$

defines a map $f\colon E \to \bar{E}$. Moreover

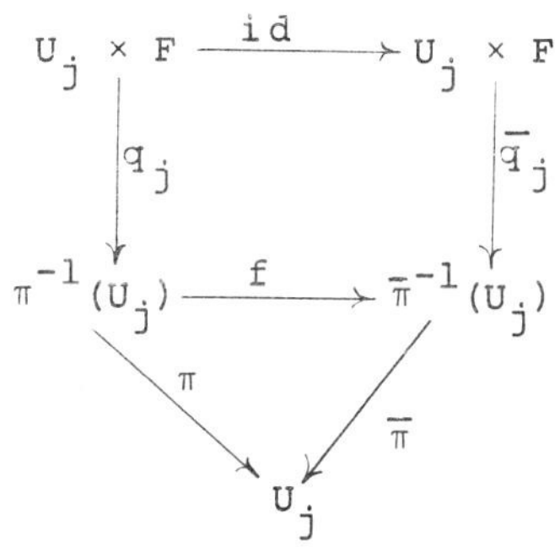


commutes, so $f$ is an equivalence of fibre bundles. Thus

<u>Theorem 2.3.4</u>. The map of Systems of Coordinate Transformations in $B$ with values in $h(F)$ to fibre bundles with fibre $F$ and base $B$ defined by $\{U_j, \phi_{ji}\} \to (F, E_{\{U_j,\phi_{ji}\}}, B, \pi)$ induces a 1-1 correspondence between equivalence classes of coordinate transformations and equivalence classes of fibre bundles.

## 4. Fibre Bundles with Structure Group

In the case of bundles over differentiable manifolds with fibre $R^n$ it seems reasonable to require that

(1) each fibre is a vector space

(2) equivalences of bundles (and maps of bundles) are isomorphisms of vector spaces when restricted to any fibre.

In terms of systems of coordinate transformations the restrictions are

(1) $\phi_{ji}: U_j \cap U_i \to h(R^n)$ factors through $GL(n,R)$

$$\text{i.e.}\quad \begin{array}{ccc} \phi_{ji}: U_j \cap U_i & \longrightarrow & h(R^n) \\ & \searrow & \uparrow \\ & & GL(n,R) \end{array}$$

(2) If the $\phi_{j',i}$ define an equivalence $\{U_j,\phi_{ji}\} \sim \{U_{j'},\phi_{j',i'}\}$ then the $\phi_{j',i}$ factor through $GL(n,R)$

$$\text{i.e.}\quad \begin{array}{ccc} \phi_{j',i}: U_{j'} \cap U_i & \longrightarrow & h(R^n) \\ & \searrow & \uparrow \\ & & GL(n,R) \end{array}$$

<u>Definition</u>. Let $G$ be a topological group, $F$ a topological space and $f: G \times F \to F$ a continuous map. Denote $f(g,x)$ by $gx$. $G$ is called a <u>topological transformation group of</u> $F$ relative to $f$ if

(1) $ex = x$ for all $x$ in $F$

(2) $(g_2g_1)x = g_2(g_1x)$.

$G$ is called effective if $gx = x$ for all $x$ in $F$ implies $g = e$.

<u>Remarks</u>. The map $F \xrightarrow{g} F$ defined by $x \to gx$ is a homeomorphism with inverse $F \xrightarrow{g^{-1}} F$.

A topological transformation group $G$ is precisely a homomorphism of groups $G \xrightarrow{f} h(F)$ such that the composition $G \times F \xrightarrow{f \times F} h(F) \times F \longrightarrow F$ is continuous.

Examples. (1) If $F$ is a locally connected, locally compact Hausdorff space, then $h(F)$ is an effective topological transformation group of $F$. In some sense it is the "largest" one. Also any topological subgroup of $h(F)$ is an effective transformation group of $F$.

(2) Let $F$ = reals, complexes, quaternions, then $GL(n,F), G(n), SO(n), SU(n)$ are effective transformation groups of $F^n$.

(3) $G(n)$ is an effective transformation group of $S^{nd-1}$.

(4) $SO(n)$ and $SU(n)$ are effective transformation groups of $S^{n-1}$ and $S^{2n-1}$ respectively.

Definition. A coordinate bundle with group $G$ consists of the following

(1) Topological spaces $F,E,B$

(2) a continuous map $\pi: E \to B$

(3) $G$-an effective topological transformation group of $F$.

(4) An indexed open cover $\{U_j\}_J$ of $B$ together with a collection $\{\phi_j\}_J$ of homeomorphisms $\phi_j: U_j \times F \to \pi^{-1}(U_j)$ satisfying

(5) $$\begin{array}{ccc} U_j \times F & \xrightarrow{\phi_j} & \pi^{-1}(U_j) \\ & {\scriptstyle P_1} \searrow & \downarrow \\ & & U_j \end{array}$$ commutes

(6) For each $u \in U_j \cap U_i$ the map $(U_j \cap U_i) \times F \xrightarrow{\phi_i} \pi^{-1}(U_j \cap U_i) \xrightarrow{\phi_j^{-1}} (U_j \cap U_i) \times F$ has the form $1 \times g$ i.e. $\phi_j^{-1}\phi_i(u,x) = (u,gx)$ for all $x$ in $F$.

(7) The map $\phi_{ji}\colon U_j \cap U_i \to G$ defined by $\phi_{ji}(u) = g$ if $\phi_j^{-1}\phi_i = 1 \times g$ is continuous.

Remark. (1) If $(F,E,B,\pi,G,\{U_j\},\{\phi_j\})$ is a coordinate bundle, then

$$\phi_{kj}(u)\,\phi_{ji}(u) = \phi_{ki}(u) \quad \text{for} \quad u \in U_k \cap U_j \cap U_i.$$

Thus each coordinate bundle with group $G$ determines a system of coordinate transformations in $B$ with values in $G$.

(2) If $F$ is a locally connected, locally compact, Hausdorff space, then "coordinate bundle" = "coordinate bundle with group $h(F)$".

The objective is to define "fibre bundle with group $G$", and "equivalences of fibre bundles with group $G$" to get a commutative diagram

Coordinate bundles with base B, fibre F, group G ⟶ Fibre bundles with base B, fibre F, group G ⟶ Equivalence classes of fibre bundles, base B, fibre F, group G

Coordinate bundles with base B, fibre F, group G ↘ Systems of coordinate transformations in B with values in G ↗ Equivalence classes of fibre bundles, base B, fibre F, group G

This diagram should reduce to the analogous diagram in Section 3 if $F$ is a locally connected, locally compact Hausdorff space and $G = h(F)$.

The bottom half of the diagram more or less defines itself. The map: Coordinate Bundles → Coordinate Transformations is provided by Remark (1) above. The obvious equivalence relation to put on the systems of coordinate transformations is

<u>Definition</u>. Let $\{U_j,\phi_{ji}\}$ and $\{U_{j'},\phi_{j',i'}\}$ be systems of coordinate transformations in $B$ with values in $G$. Set $\{U_j,\phi_{ji}\} \sim \{U_{j'},\phi_{j',i'}\}$ iff there exist maps

$$\phi_{j'i}\colon U_{j'} \cap U_i \to G \quad \text{satisfying}$$

equation 1. $\phi_{k'i} = \phi_{k'j'}\phi_{j'i}$ on $U_{k'} \cap U_{j'} \cap U_i$

equation 2. $\phi_{k'i} = \phi_{k'j}\phi_{ji}$ on $U_{k'} \cap U_j \cap U_i$

Remarks. (1) $\{U_j,\phi_{ji}\} \sim \{U_{j'},\phi_{j'i'}\}$ iff $\phi_{j'i}$ and $\phi_{ij'}$ can be defined so that $(\{U_j\} \cup \{U_{j'}\}, \phi_{ji},\phi_{j'i'},\phi_{j',i},\phi_{ij'})$ forms a system of coordinate transformations in B with values in G.

(2) $\sim$ is an equivalence relation (see Section 3).

So we have

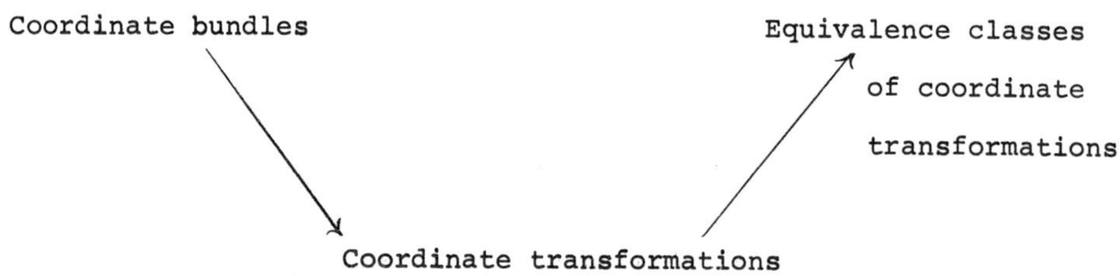


We shall define "Fibre Bundle with group G" as an equivalence class of coordinate bundles with group G. First, note that if "Fibre Bundle with group G" is to reduce to the old notion of fibre bundle when $G = h(F)$, then two coordinate bundles can not represent the same fibre bundle with group G unless they have the same total space and projection. Thus the relation must have the form $(F,E,B,\pi,G,\{U_j\},\{\phi_j\})$ is equivalent to $(F,E,B,\pi,G,\{U_{j'}\},\{\phi_{j'}\})$ iff --- . On the other hand we cannot simply "forget" the $\{U_j\}$, $\{\phi_j\}$ because we want a well defined map from the set of fibre bundles with group G into the equivalence classes of coordinate transformations with group G.

Definition. The coordinate bundles $(F,E,B,\pi,G,\{U_j\},\{\phi_j\})$ and $(F,E,B,\pi,G,\{U_{j'}\},\{\phi_{j'}\})$ are called strictly equivalent if

(1) for each $u \in U_{j'} \cap U_i$ the map $\phi_{j'}^{-1}\phi_i$: $(U_{j'} \cap U_i) \times F \to (U_j \cap U_i) \times F$ has the form $(u,x) \to (u,gx)$

and

(2) the map $\phi_{j'i}$: $U_{j'} \cap U_i \to G$ defined by $\phi_{j'i}(u) = g$ if $\phi_{j'}^{-1}\phi_i(u,x) = u,gx)$ for all $x$ in $F$, is continuous.

Notation: $(F,E,B,\pi,G,\{U_j\},\{\phi_j\}) \overset{S}{\sim} (F,E,B,\pi,G,\{U_{j'}\},\{\phi_{j'}\})$

Remarks. (1) $\overset{S}{\sim}$ is an equivalence relation (see previous section)

(2) $(F,E,B,\pi,G,\{U_j\},\{\phi_j\}) \overset{S}{\sim} (F,E,B,\pi,G,\{U_{j'}\},\{\phi_{j'}\})$ iff $(F,E,B,\pi,G,\{U_j\} \cup \{U_{j'}\}, \{\phi_j\} \cup \{\phi_{j'}\}$ is a coordinate bundle.

Definition. A fibre bundle with base B, fibre F, and group G is by definition an equivalence class of coordinate bundles. The fibre bundle determined by $(F,E,B,\pi,G,\{U_j\},\{\phi_j\})$ is denoted by $(F,E,B,\pi,G,[\{U_j\},\{\phi_j\}])$.

Remarks. (1) if $F$ is locally connected, locally compact, Hausdorff and $G = h(F)$, then any two coordinate bundles of the form $(F,E,B,\pi,h(F),\{U_j\},\{\phi_j\})$, $(F,E,B,\pi,h(F),\{U_{j'}\},\{\phi_{j'}\})$ are strictly equivalent. So "fibre bundle with group $h(F)$" = "fibre bundle".

(2) In view of the preceeding Remark (2), if two coordinate bundles with group G are strictly equivalent, then their corresponding systems of coordinate transformations are equivalent. Hence we have a commutative diagram

Set of coordinate bundles with base B, fibre F, group G $\longrightarrow$ Set of fibre bundles with base B, fibre F, group G $\longrightarrow$ Set of equivalence classes of systems of coordinate transformations

Set of coordinate bundles with base B, fibre F, group G $\searrow$ Set of systems of coordinate transformations $\nearrow$ Set of equivalence classes of systems of coordinate transformations

It remains to define an equivalence relation on the set of fibre bundles so that the set of equivalence classes of fibre bundles is in a 1-1 correspondence with the set of equivalence classes of systems of coordinate transformations.

If the coordinate bundles $(F,E,B,\pi,G,\{U_j\},\{\phi_j\})$ and $(F,E',B,\pi',G,\{U_{j'}\},\{\phi_{j'}\})$ are to determine equivalent fibre bundles then there should be a homeomorphism $E \xrightarrow{f} E'$ which is an equivalence of bundles in the old sense. Also $\{U_{j'},\phi_{j',i'}\}$, $\{U_j,\phi_{ji}\}$ should be equivalent systems of coordinate transformations. This motivates

Definition. Two coordinate bundles $(F,E,B,\pi,G,\{U_j\},\{\phi_j\})$ and $(F,E',B,\pi',G,\{U_{j'}\},\{\phi_{j'}\})$ are called equivalent if there is an equivalence $f: E' \to E$ of bundles in the sense of Chapter 2, section 1, such that $(F,E,B,\pi,G,\{U_j\},\{\phi_j\})$ and $(F,E,B,\pi,G,\{U_{j'}\},\{f\circ\phi_{j'}\})$ are strictly equivalent coordinate bundles.

Remarks. (a) It is easy to see that if $f: E' \to E$ is an equivalence of bundles in the old sense, then $(F,E,B,\pi,G,\{U_{j'}\},\{f\circ\phi_{j'}\})$ is a coordinate bundle with group G. Moreover, $(F,E,B,\pi,G,\{U_{j'}\},\{f\circ\phi_{j'}\})$ and $(F,E',B,\pi',G,\{U_{j'}\},\{\phi_{j'}\})$ determine the same system of coordinate transformations. So equivalent coordinate bundles determine equivalent systems of coordinate transformations.

(b) Equivalence of coordinate bundles is an equivalence relation.

(c) Strictly equivalent coordinate bundles are equivalent. Thus equivalence of coordinate bundles can be viewed as a relation on Fibre bundles. We have the commutative diagram

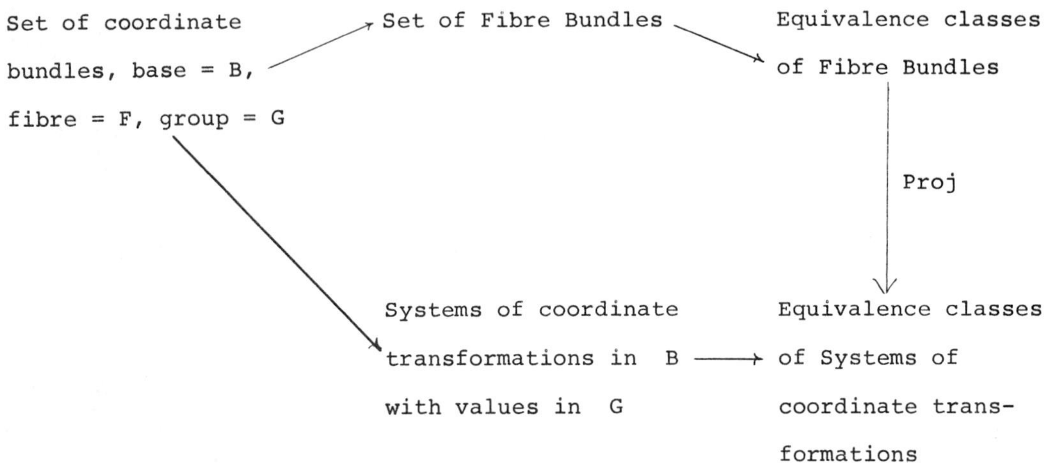


To show that (proj) is 1-1 and onto it suffices to show that there is a map $\rho$: "Equivalence classes of coordinate transformations" $\to$ "Equivalence classes of Fibre Bundles" such that $(\mathrm{Proj})\circ f = \mathrm{id}$ and $f\circ \mathrm{proj} = \mathrm{id}$. The arguments of the previous section are easily generalized to cover this situation.

## 5. Constructions and the Bundle Structure Theorem for Homogeneous Spaces

In this section we tie up some loose ends before preceeding to the covering homotopy theorems.

Definition. A map $(F,E',B,\pi',G,\{U_{j'}\},\{\phi_{j'}\}) \to (F,E,B,\pi,G,\{U_j\},\{\phi_{j'}\})$ of coordinate bundles is a commutative diagram

$$\begin{array}{ccc} E' & \xrightarrow{\tilde{f}} & E \\ \downarrow{\scriptstyle \pi'} & & \downarrow{\scriptstyle \pi} \\ B' & \xrightarrow{f} & B \end{array}$$

such that for each $u \in U_{j'} \cap f^{-1}(U_i)$ the map $\phi_i^{-1}\tilde{f}\phi_{j'}: F \to F$ defined by

$$\begin{array}{ccc} \{u\} \times F & & \{f(u)\} \times F \\ \downarrow{\scriptstyle \phi_{j'}} & & \uparrow{\scriptstyle \phi_i^{-1}} \\ \pi'^{-1}\{u\} & \xrightarrow{\tilde{f}} & \pi^{-1}\{f(u)\} \end{array}$$

is given by $x \to gx$ for some $g \in G$, and that the corresponding map $\phi_i^{-1}\tilde{f}\phi_{j'}: U_{j'} \cap f^{-1}(U_i) \to G$ is continuous.

It is readily seen that the identity map $E \to E$ is a map of coordinate bundles. Likewise the composition of maps of coordinate bundles is a map of coordinate bundles.

Let $(F,E,B,\pi,G,\{U_j\},\{\phi_j\})$ be a coordinate bundle and $f: B' \to B$ a continuous map. Set $E'$ equal to the subspace of $B' \times E$ consisting of pairs $(b',x)$ such that $f(b') = \pi(x)$. Define $\tilde{f}: E' \to E$ by $\tilde{f}(b',x) = x$ and $\pi': E' \to B'$ by $\pi'(b',x) = b'$. From Lemma 2.2.2, we know that $E'$ is a bundle (in the sense of Section 1) over $B'$ with coordinate nbhds $\{f^{-1}(U_j)\}$ and coordinate functions $\overline{\phi}_j: f^{-1}(U_j) \times F \to \pi'^{-1}(f^{-1}(U_j))$ defined by $\overline{\phi}_j(u,x) = (u,\phi_j(f(u),x))$. It follows that $(F,E',B',\pi',G,\{f^{-1}(U_j)\},\{\overline{\phi}_j\})$ is a coordinate bundle. $(F,E',B'\pi',G,\{f^{-1}(U_j)\},\{\overline{\phi}_j\})$ is called the <u>pullback of</u> $(F,E,B,\pi,G,\{U_j\},\{\phi_j\})$ <u>over</u> $f$

The following are left as exercises for the reader

1. The map $\tilde{f}\colon E' \to E$ is a map of coordinate bundles.

2. The system of coordinate transformations determined by the bundle $(F,E',B',\pi',G,\{f^{-1}(U_j)\},\{\bar{\phi}_j\})$ is given by

$$\begin{array}{ccc} f^{-1}(U_j) \cap f^{-1}(U_i) & \xrightarrow{\ \bar{\phi}_{ji}\ } & G \\ & {\scriptstyle f}\searrow \qquad \nearrow{\scriptstyle \phi_{ji}} & \\ & U_j \cap U_i & \end{array}$$

3. Let $f\colon B' \to B$ be continuous, the pullback construction induces a well defined map

$$\begin{array}{ccc} \text{Set of equivalence} & & \text{Set of equivalence} \\ \text{classes of fibre} & \xrightarrow{\quad f^{*}\quad} & \text{classes of fibre} \\ \text{bundles with base } B, & & \text{bundles with base } B', \\ \text{fibre } F, \text{ group } G & & \text{fibre } F, \text{ group } G \end{array}$$

4. Let $(F,\bar{E},B',\bar{\pi},G,\{U_{j'}\},\{\phi_{j'}\})$ be a coordinate bundle and let

$$\begin{array}{ccc} \bar{E} & \xrightarrow{\ \bar{f}\ } & E \\ \downarrow{\scriptstyle \bar{\pi}} & & \downarrow{\scriptstyle \pi} \\ B' & \xrightarrow{\ f\ } & B \end{array}$$

be commutative. Then $\bar{f}$ is a map of coordinate bundles iff the map $\bar{E} \xrightarrow{h} E'$ defined by

$h(y) = (\bar{\pi}y, \bar{f}(y))$ is an equivalence of coordinate bundles over $B'$, with group $G$.

5. Two coordinate bundles $(F,E,B,\pi,G,\{U_j\},\{\phi_j\})$ and $(F,E',B,\pi',G,\{U_{j'}\},\{\phi_{j'}\})$ are equivalent iff there is a map of coordinate bundles with group $G$: $f'\colon E' \to E$ so that

$$\begin{array}{ccc} E' & \xrightarrow{f'} & E \\ \downarrow{\scriptstyle \pi'} & & \downarrow{\scriptstyle \pi} \\ B & \xrightarrow[1_B]{} & B \end{array}$$

commutes.

<u>Definition</u>. Let $(F,E,B,\pi,G,\{U_j\},\{\phi_j\})$ and $(F',E',B',\pi',G',\{U_{j'}\},\{\phi_{j'}\})$ be coordinate bundles. Then $(F \times F', E \times E', B \times B', \pi \times \pi', G \times G', \{U_j \times U_{j'}\}, \{\phi_{(j,j')}\})$ is a coordinate bundle called the <u>product of</u> $E$ <u>and</u> $E'$ where $\phi_{(j,j')}\colon (U_j \times U_{j'}) \times F \times F' \to (\pi \times \pi')^{-1}(U_j \times U_{j'})$ is defined by $\phi_{(j,j)}\colon ((u,u'),(x,x')) = (\phi_j(u,x),\phi_{j'}(u',x'))$ in $E \times E'$. The corresponding system of coordinate transformations $\{U_j \times U_{j'}, \phi_{(j,j')(i,i')}\}$ is given by

$$\begin{array}{ccc} (U_j \times U_{j'}) \cap (U_i \times U_{i'}) & \xrightarrow{\phi_{(j,j')(i,i')}} & G \times G' \\ & {\scriptstyle id}\searrow \qquad\qquad \nearrow{\scriptstyle \phi_{ji} \times \phi_{j'i'}} & \\ & (U_j \cap U_i) \times (U_{j'} \cap U_{i'}) & \end{array}$$

Definition. A coordinate bundle $(F,E,B,\pi,G,\{U_j\},\{\phi_j\})$ is called principal if $F = G$, and the action of $G$ on $F$ is given by the multiplication in $G$. By passing to systems of coordinate transformations and using the construction of the previous section we see that any coordinate bundle $(F,E,B,\pi,G,\{U_j\},\{\phi_j\})$ determines a principal bundle $(G,E',B,\pi',G,\{U_j\},\{\phi_{j'}\})$ called the associated principal bundle of $E$, and that this map on the set of coordinate bundles induces a 1-1, onto map on the set of equivalence classes of fibre bundles. Often the principal bundle is simpler than the original bundle. For example, the Mobius Strip, Klein Bottle, and twisted torus all have as principal bundle the double covering of S'.

More generally, we say that two bundles are associated if their principal bundles are equivalent. So the Mobius Strip, Klein Bottle, and twisted torus are associated $Z_2$ bundles. It is not hard to see that if two bundles are equivalent then they are associated. Moreover, if two associated bundles have the same fibre, and the same action of the group on the fibre, then they are equivalent.

Note: Twisted Torus and Klein Bottle are associated $Z_2$ bundles. They are not equivalent as $Z_2$ bundles since the actions of $Z_2$ on $S^1$ are different.

In this way all bundles over $B$ with group $G$ are divided into non-overlapping classes; within each class is just one equivalence class of principal bundles.

If $H$ is a subgroup of $G$, and $G$ acts effectively on $F$ then any coordinate bundle with fibre $F$ and group $H$ determines a coordinate bundle with fibre F and group G, called the G image of the original coordinate bundle.

For example the action of $Z_2$ on S for the twisted torus is given by $a(\theta) = \pi + \theta$. Map $Z_2 \to S$ by $a \to \pi$ and define $S^1 \times S^1 \to S^1$ by $\alpha(\theta) = \alpha + \theta$. In this way the $Z_2$ action on $S^1$ extends to an $S^1$ action on $S^1$. The $S^1$ image of the twisted torus is a principal $S^1$ bundle which is equivalent to the trivial bundle.

In general, given a coordinate bundle $(F,E,B,\pi,G,\{U_j\},\{\phi_j\})$ one would like to reduce the group, i.e. find a bundle $(F,E',B,\pi',H,\{U_{j'}\},\{\phi_{j'}\})$ where $H \xrightarrow{1-1} G$ so that the G-image of E' is equivalent to E. The idea being: the smaller the group, the simpler the bundle. Note: a bundle is equivalent to the trivial bundle iff the group of the bundle can be reduced to the trivial group e.

Definition. If B is a topological space and G a topological group, set $K_G(B)$ equal to the set of equivalence classes of principal bundles with base B and group G. Then $K_G(B)$ is a set with distinguished element. The distinguished element being the element of $K_G(B)$ corresponding to the trivial bundle.

The following statements are immediate

1. Given a map $f: B' \to B$ the pullback construction induces a map $f^*: K_G(B) \to K_G(B')$ of sets with distinguished element.

2. A continuous homomorphism $h: H \to G$ of topological groups induces a map $h_*: K_H(B) \to K_G(B)$ of sets with distinguished element.

3. The product construction induces a map $K_G(B) \times K_{G'}(B') \longrightarrow K_{G\times G'}(B \times B')$ of sets with distinguished element. Note that the map $K_G(B) \times K_{G'}(B) \to K_{G\times G'}(B)$ induced by the Whitney Sum construction is the composition

$$K_G(B) \times K_{G'}(B) \xrightarrow{\times} K_{G\times G'}(B \times B) \xrightarrow{\Delta^*} K_{G\times G'}(B).$$

4. If $G$ is an abelian topological group with multiplication $u_G: G \times G \to G$, then the composition

$$K_G(B) \times K_G(B) \xrightarrow{\times} K_{G\times G}(B \times B) \xrightarrow{\Delta^*} K_{G\times G}(B) \xrightarrow{(u_G)_*} K_G(B)$$

gives $K_G(B)$ the structure of an abelian group. The identity of $K_G(B)$ being the element of $K_G(B)$ determined by the trivial bundle.

<u>Theorem 2.5.1</u>. Let $(G,E,B,\pi,G,\{U_j\},\{\phi_j\})$ be a principal bundle. Then there is a map $h: E \times G \to E$ such that

$$\text{(1)} \quad \begin{array}{ccc} E \times G \times G & \xrightarrow{1_E \times u_G} & E \times G \\ \downarrow{\scriptstyle h\times 1_G} & & \downarrow{\scriptstyle h} \\ E \times G & \xrightarrow{h} & E \end{array} \quad \text{commutes}$$

and

$$\begin{array}{ccc} E \times G & \xrightarrow{\;h\;} & E \\ \big\downarrow{\scriptstyle p_1} & & \big\downarrow{\scriptstyle \pi} \\ E & \xrightarrow{\;\pi\;} & B \end{array}$$

is a map of coordinate bundles, where $p_1$ is projection onto the first factor and $E \times G \xrightarrow{p_1} E$ is given the structure of the trivial G-bundle over $E$. $h$ is called the principal map of the bundle $E$.

Proof. Define the map $h: E \times G \to E$ on $\pi^{-1}(U_i)$ by requiring that

$$\begin{array}{ccc} (U_i \times G) \times G & \xrightarrow{\;1_{U_i} \times u_G\;} & U_i \times G \\ \big\downarrow{\scriptstyle \phi_i \times 1_G} & & \big\downarrow{\scriptstyle \phi_i} \\ \pi^{-1}(U_i) \times G & \xrightarrow{\;h\;} & \pi^{-1}(U_i) \end{array}$$

be commutative. $h$ is well defined since

$$\begin{array}{ccccccc} & \overset{\phi_i^{-1} \times 1_G}{\nearrow} & (U_i \cap U_j) \times G \times G & \xrightarrow{\;1 \times u_G\;} & (U_i \cap U_j) \times G & \overset{\phi_i}{\searrow} & \\ \pi^{-1}(U_i \cap U_j) \times G & & & & & & \pi^{-1}(U_i \cap U_j) \\ & \underset{\phi_j^{-1} \times 1_G}{\searrow} & (U_i \cap U_j) \times G \times G & \xrightarrow{\;1 \times u_G\;} & (U_i \cap U_j) \times G & \underset{\phi_j}{\nearrow} & \end{array}$$

commutes

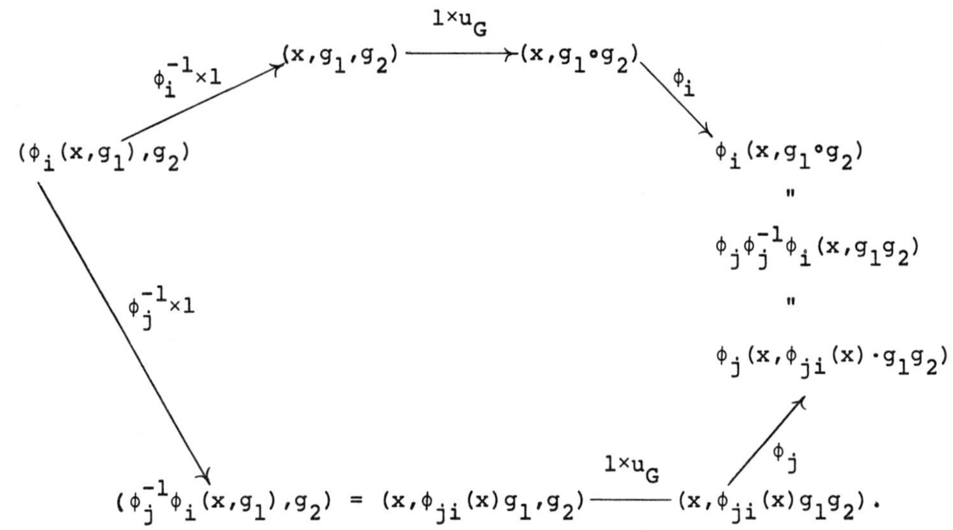


The commutivity of diagram (1) follows from the definition of $h$. To show that $h\colon E \times G \to E$ is a map of coordinate bundles, it suffices to show that for each $U_i$ and $u$ in $\pi^{-1}(U_i)$, the map

$\mu \times G \xrightarrow{h} \pi^{-1}(\pi(u)) \xrightarrow{\phi_i^{-1}} \pi(u) \times G$ has the form $(u,g) \to (\pi(u), g_u g)$

and that the corresponding map $\pi^{-1}(U_i) \to G$ defined by $\mu \to g_u$ is continuous. From the definition of $h$, it follows that

$$\mu \times G \xrightarrow{h} \pi^{-1}(\pi(u)) \xrightarrow{\phi_i^{-1}} \pi(u) \times G \text{ is given by}$$

$(u,g) \to (\pi(u), P_2\phi_i^{-1}(u)g)$ where $P_2\phi_i^{-1}$ is the composition

$\pi^{-1}(U_i) \xrightarrow{\phi_i^{-1}} U_i \times G \xrightarrow{P_2} G.$ The result follows.

If $E \xrightarrow{\pi} B$ is a principal G-bundle and G acts effectively on the space F, set $E \times_G F$ equal to the quotient space of $E \times F$ obtained by identifying $(xg,y)$ with $(x,gy)$. Define $E \times_G F \xrightarrow{\pi'} B$ by $[x,y] \to \pi(x)$, then $E \times_G F \xrightarrow{\pi'} B$ can be given the structure of a coordinate bundle with group G and fibre F such that $E \xrightarrow{\pi} B$ is equivalent to the associated principal bundle of $E \times_G F \xrightarrow{\pi'} B$.

Let G be a topological group and let H be a closed subgroup of G. Then G acts as a group of homeomorphisms on the homogeneous space G/H. Set $H_o = \{g \in G\colon g\colon G/H \to G/H$ is the identity$\}$. It follows that $H_o = \bigcap_{g \in G} gHg^{-1}$. Since H is closed, $gHg^{-1}$ is closed for all g in G. So $H_o$ is a closed normal subgroup of G, and $G/H_o$ acts effectively on G/H as a group of homeomorphisms. If N is a subgroup of H, and a normal subgroup of G, then $N \subseteq H_o$. $H_o$ is called the largest subgroup of H which is a normal subgroup of G.

<u>Definition</u>. Let G be a closed subgroup of the topological group T. A local section of G in T is a continuous map $f\colon U \to T$, (where U is a nbhd of eG in T/G) such that $\pi \circ f = 1_U$, $\pi\colon T \to T/G$.

<u>Theorem 2.5.2</u>. <u>Bundle Structure Theorem for Homogeneous Spaces</u>. Let $H \subset G \subset T$ be inclusions of topological groups, with H a closed subgroup of G, and G a closed subgroup of T. Let $\pi\colon T/H \to T/G$ be the projection of cosets. If G admits a local section in T, then we can assign a bundle structure to $\pi\colon T/H \to T/G$. The fibre

is $G/H$, and the group is $G/H_o$ acting on $G/H$ as left translations where $H_o$ is the largest subgroup of $H$ which is a normal subgroup of $G$. Moreover, any two local sections lead to strictly equivalent bundles.

Proof.

Let $f: \underset{\substack{T/G}}{U} \longrightarrow T$ be a local section. (Here $U \subset T/G$.)

Note: $f(tG)G = tG$ in $T/G$ and so $[f(tG)]^{-1}t \in G$ for all $t \in T$.

Define

$$U \times G/H \xrightarrow{\phi_e} \pi^{-1}(U) \xrightarrow{\phi_e^{-1}} U \times G/H \qquad G/H \qquad \begin{array}{c} T/H \\ \downarrow \pi \\ T/G \end{array}$$

by

$$(tG, gH) \xrightarrow{\phi_e} f(tG)gH$$

and

$$tH \xrightarrow{\phi_e^{-1}} (tG, [f(tG)]^{-1}tH).$$

Note: $\phi_e^{-1}(f(tG), gH) = (tG, [f(tG)]^{-1}f(tG)gH) = (tG, gH)$.

For each $t \in T$, set $U_t = tU = \{tt'G\colon t'G \in U\} \subseteq T/G$. Since $T$ acts transitively on $T/G$, $\{U_t\}_{t \in G}$ is an open covering of $T/G$. Define

$$U_t \times G/H \xrightarrow{\phi_t} \pi^{-1}(U_t) \xrightarrow{\phi_t^{-1}} U_t \times G/H$$

by

$$\begin{array}{ccccc}
U_t \times G/H & \xrightarrow{\phi_t} & \pi^{-1}(U_t) & \xrightarrow{\phi_t^{-1}} & U_t \times G/H \\
\downarrow {\scriptstyle t^{-1}\times 1} & & {\scriptstyle t}\uparrow \quad \downarrow {\scriptstyle t^{-1}} & & \uparrow {\scriptstyle t\times 1} \\
U \times G/H & \xrightarrow{\phi_e} & \pi^{-1}(U) & \xrightarrow{\phi_e^{-1}} & U \times G/H
\end{array}$$

All maps are homeomorphisms: By definition of $\phi_t$ and $\phi_t^{-1}$ we see that

$$\phi_t^{-1} \circ \phi_t = (t\times 1)\circ(\phi_e^{-1})\circ(\phi_e)\circ(t^{-1}\times 1) = (t\times 1)(t^{-1}\times 1) = 1 \times 1 = \text{identity}.$$

Next we want to calculate

$$(U_t \cap U_{t_o}) \times G/H \xrightarrow{\phi_{t_o}} \pi^{-1}(U_t \cap U_{t_o}) \xrightarrow{\phi_t^{-1}} (U_t \cap U_{t_o}) \times G/H$$

$$\begin{array}{ccccccc}
(U_t \cap U_{t_o}) \times G/H & \xrightarrow{\phi_{t_o}} & \pi^{-1}(U_t \cap U_{t_o}) & & \xrightarrow{\phi_t^{-1}} & & (U_t \cap U_{t_o}) \times G/H \\
\downarrow t_o^{-1} \times 1 & & \nearrow t_o & \searrow t^{-1} & & & \uparrow t \times 1 \\
(U_{t_o^{-1}t} \cap U) \times G/H & \xrightarrow{\phi_e} & \pi^{-1}(U_{t_o^{-1}t} \cap U) & \xrightarrow{t^{-1}t_o} & \pi^{-1}(U \cap U_{t^{-1}t_o}) & \xrightarrow{\phi_e^{-1}} & (U \cap U_{t^{-1}t_o}) \times G/H
\end{array}$$

$$\begin{array}{ccccccc}
(\hat{t}G, gH) & & \xrightarrow{\phi_t^{-1} \circ \phi_{t_o}} & & & & (\hat{t}G, [f(t'\hat{t}G)]^{-1}t^{-1}\hat{t}_o f(t_o^{-1}\hat{t}G)gH) \\
\downarrow t_o^{-1} \times 1 & & & & & & \uparrow t \times 1 \\
(\hat{t}_o^{-1}\hat{t}G, gH) & \xrightarrow{\phi_e} & f(t_o^{-1}\hat{t}G)gH & \xrightarrow{t^{-1}t_o} & t^{-1}t_o f(t_o^{-1}\hat{t}G)gH & \xrightarrow{\phi_e^{-1}} & (t^{-1}\hat{t}G, [f(t^{-1}\hat{t}G)]^{-1}t^{-1}t_o f(t_o^{-1}\hat{t}G)gH).
\end{array}$$

Note that $[f(t^{-1}\hat{t}G)]^{-1}t^{-1}t_o f(t_o^{-1}\hat{t}G)$ is an element of $G$ since $f(t^{-1}\hat{t}G) = f[t^{-1}t_o f(t_o^{-1}\hat{t}G)G]$ and $[f(tG)]^{-1}t \in G$ for all $t \in T$.

For $t$, and $t_o$ in $T$ define $\phi_{t,t_o}: (U_t \cap U_{t_o}) \to G/H_o$ by $\hat{t}G \to [f(t^{-1}\hat{t}G)]^{-1}t^{-1}t_o f(t_o^{-1}\hat{t}G)H_o$. Then $\phi_{t,t_o}$ is continuous and $(U_t \cap U_{t_o}) \times G/H \xrightarrow{\phi_t^{-1}\phi_{t_o}} (U_t \cap U_{t_o}) \times G/H$ is given by $(\hat{t}G,gH) \to (\hat{t}G,\phi_{t,t_o}(\hat{t}G)gH)$.

Thus $(G/H,T/H,T/G,\pi,G/H_o,\{U_t\}_{t\epsilon T},\{\phi_t\})$ is a coordinate bundle. The corresponding system of coordinate transformations is $\{U_t,\phi_{t,t_o}\}$.

Next we want to show that different cross sections give rise to strictly equivalent coordinate bundles. Recall that two coordinate bundles $(F,E,B,\pi,G,\{U_j\},\{\phi_j\})$ and $(F,E,B,\pi,\{U_j'\},\{\phi_j'\})$ are strictly equivalent iff

$$(U_k \cap U_j') \times F \xrightarrow{\phi_k} \pi^{-1}(U_k \cap U_j') \xrightarrow{(\phi_j')^{-1}} (U_k \cap U_j') \times F$$

coincides with the operation of an element in $G$, and the corresponding map:

$$\overline{\phi}_{j',k}: (U_k \cap U_j') \to G \quad \text{is continuous.}$$

Let $U'$ be a nbhd of $eG$ in $T/G$ and let $f': U' \to T/H$ be a local cross section. Define $\phi_t': U_t' \times G/H \to \pi^{-1}(U_t') \xrightarrow{(\phi_t')^{-1}} U_t' \times G/H$ by

$$\phi_t'(\hat{t}G, gH) = tf'(t^{-1}\hat{t}G)gH$$

and

$$(\phi_t')^{-1}(\hat{t}H) = (\hat{t}G, [f'(t^{-1}\hat{t}G)]^{-1}t^{-1}\hat{t}H).$$

Then as above $(G/H, T/H, T/G, \pi, G/H_o, \{U_t'\}_{t\varepsilon T}, \{\phi_t'\})$ is a coordinate bundle. To check that $(G/H, T/H, T/G, \pi, G/H_o, \{U_t'\}, \{\phi_t'\})$ is strictly equivalent to $(G/H, T/H, T/G, \pi, G/H_o, \{U_t\}, \{\phi_t\})$ consider

$$(U_t \cap U_{t_o}') \times G/H \xrightarrow{\phi_t} \pi^{-1}(U_t \cap U_{t_o}') \xrightarrow{(\phi_{t_o}')^{-1}} (U_t \cap U_{t_o}') \times G/H$$

$$(\hat{t}G, gH) \xrightarrow{\phi_t} tf(t^{-1}\hat{t}G)gH \xrightarrow{(\phi_{t_o}')^{-1}} (\hat{t}G, [f'(t_o^{-1}\hat{t}G)]^{-1}t_o^{-1}tf(t^{-1}\hat{t}G)gH).$$

So $\overline{\phi}_{t_o,t}: (U_t \cap U_{t_o}') \to G/H_o$ is given by

$$\hat{t}G \to [f'(t_o^{-1}\hat{t}G)]^{-1}t_o^{-1}tf(t^{-1}\hat{t}G)$$

which is continuous. Q.E.D.

Corollary 2.5.3. Let $H \subset G \subset T$ be inclusions of topological groups with $H$ a closed subgroup of $G$ and $G$ a closed subgroup of $T$. Assume $G$ admits a local section in $T$. Then the principal bundle associated with the bundle

$$\begin{array}{ccc} G/H & \longrightarrow & T/H \\ & & \downarrow \\ & & T/G \end{array}$$

is the bundle

$$\begin{array}{ccc} G/H_o & \longrightarrow & T/H_o \\ & & \downarrow \\ & & T/G \end{array},$$

where $H_o$ is the largest subgroup of $H$ which is a normal subgroup of $G$.

Proof. From the proof of Theorem 2.5.1 we see that a local section

$$f: U \to T, \ (U \subseteq T/G),$$

determines the same system of coordinate transformations for the two bundles above.

Corollary 2.5.4. Let $H$ be a closed subgroup of $G$. Assume $H$ admits a local section in $G$. Then

$$\begin{array}{ccc} H & \longrightarrow & G \\ & & \downarrow \\ & & G/H \end{array}$$

is a principal fibre bundle.

Proof. This is a special case of Theorem 2.5.2.

Examples: For $k < n$, $G(k)$ admits a local section in $G(n)$, and $G(k) \times G(n-k)$ admits a local section in $G(n)$. (Recall $G(n,k)$ denotes $G(n)/G(k) \times I$, and $M(n,k)$ denotes $G(n)/G(k) \times G(n-k)$).

Using the Bundle Structure Theorem for Homogeneous Spaces, we see that

1. $\begin{array}{c} G(n) \\ \downarrow \pi \\ G(n,k) \end{array}$ is a principal $G(k)$ bundle. The principal map $G(n) \times G(k) \to G(n)$ is given by $(A,B) \to A\begin{pmatrix} B & 0 \\ 0 & I \end{pmatrix}$

2. $\begin{array}{c} G(n,\ell) \\ \downarrow \\ G(n,k) \end{array}$ is a $G(k)$ bundle with fibre $G(k,\ell)$ for $0 < \ell < k < n$. The associated principal bundle is equivalent to 1 above.

3. $\begin{array}{c} G(n) \\ \downarrow \\ M(n,k) \end{array}$ is a principal $G(k) \times G(n-k)$ bundle. The principal map $G(n) \times (G(k) \times G(n-k)) \to G(n)$ is given by $\left(A, \begin{pmatrix} B & 0 \\ 0 & C \end{pmatrix}\right) \longrightarrow A\begin{pmatrix} B & 0 \\ 0 & C \end{pmatrix}$

4. $\begin{array}{c} G(n,k) \\ \downarrow \\ M(n,k) \end{array}$ is a principal $G(n-k)$ bundle. The elements of $G(n,k)$ can be represented by $n \times (n-k)$ matrices $A$ whose columns are orthonormal. The principal map $G(n,k) \times G(n-k) \to G(n,k)$ is then given by $(A,B) \to AB$.

5. We get the following sequence of bundles

$$\begin{array}{rcl}
S^{d-1} = G(1) & \to & G(n) \\
 & & \downarrow \\
S^{2d-1} = G(2,1) & \to & G(n,1) \\
 & & \downarrow \\
S^{3d-1} = G(3,2) & \to & G(n,2) \\
 & & \downarrow \\
 & & \vdots \\
 & & \downarrow \\
S^{(i+1)d-1} = G(i+1,i) & \to & G(n,i) \\
 & & \downarrow \\
 & & G(n,i+1) \\
 & & \downarrow \\
 & & \cdot \\
S^{(n-1)d-1} = G(n-1,n-2) & \to & G(n,n-2) \\
 & & \downarrow \\
 & & G(n,n-1) = S^{nd-1}.
\end{array}$$

6. The commutative diagram

$$\begin{array}{ccccc}
 & & & (A,B) \to \begin{pmatrix} A & 0 \\ 0 & B \end{pmatrix} & \\
I \times G(k) & \longrightarrow & G(n) \times G(k) & \longrightarrow & G(n+k) \\
\Big\downarrow \begin{array}{c} (A) \\ \downarrow \\ \begin{pmatrix} A & 0 \\ 0 & 1 \end{pmatrix} \end{array} & & \Big\downarrow \begin{array}{c} (A,B) \\ \downarrow \\ \left[A, \begin{pmatrix} B & 0 \\ 0 & 1 \end{pmatrix}\right] \end{array} & & \Big\downarrow \begin{array}{c} A \\ \downarrow \\ \begin{pmatrix} A & 0 \\ 0 & 1 \end{pmatrix} \end{array} \\
I \times G(k+1) & \longrightarrow & G(n) \times G(k+1) & \longrightarrow & G(n+k+1)
\end{array}$$

induces maps of principal $G(n)$ bundles:

$$\begin{array}{ccccccc}
G(n+k,k) & \longrightarrow & G(n+k+1,k+1) & \longrightarrow & G(n+k+2,k+2) & \longrightarrow & \cdots \\
\downarrow & & \downarrow & & \downarrow & & \\
M(n+k,k) & \longrightarrow & M(n+k+1,k+1) & \longrightarrow & M(n+k+2,k+2) & \longrightarrow & \cdots
\end{array}$$

$$\begin{array}{ccc}
(A) & \longrightarrow & \begin{pmatrix} A \\ 0 \end{pmatrix} \\
\uparrow & & \\
(n+k) \times n \text{ matrix} & &
\end{array}$$

7. Set $E(n+k,k)$ equal to the subspace of $M(n+k,k) \times F^{n+k}$ consisting of pairs $(P,v)$ such that the vector $v$ is in the plane $P$. Then $E(n+k,k)$ can be given the

$$\begin{array}{c} E(n+k,k) \\ \Big\downarrow \; \begin{matrix} (P,v) \\ \downarrow \\ P \end{matrix} \\ M(n+k,k) \end{array}$$

structure of a $G(n)$ bundle with fibre $F^n$ such that the associated principal bundle is equivalent to the bundle $\begin{array}{c} G(n+k,k) \\ \downarrow \pi \\ M(n+k,k) \end{array}$. In particular, $E(n+k,k)$ can be identified with $G(n+k,k) \underset{G(n)}{\times} F^n$. This gives maps of coordinate bundles, with fibre $F^n$, and group $G(n)$.

$$\begin{array}{ccccccccc}
E(n+k,k) & \longrightarrow & E(n+k+1,k+1) & \longrightarrow --- \longrightarrow & E(n+r,r) & \longrightarrow & E(n+r+1,r+1) & \longrightarrow --- \\
\downarrow & & \downarrow & & \downarrow & & \downarrow & \\
M(n+k,k) & \longrightarrow & M(n+k+1,k+1) & \longrightarrow --- \longrightarrow & M(n+r,r) & \longrightarrow & M(n+r+1,r+1) & \longrightarrow ---
\end{array}$$

<u>Exercises</u>

1. For $k < n$ find a local section of $G(k)$ in $G(n)$.

2. For $k < n$ find a local section of $G(k) \times G(n-k)$ in $G(n)$.

3. Find an embedding of $E(n+k,k)$ in $M(n+k+1,k+1)$ so that the embedding of the zero section of $E(n+k,k)$ in $M(n+k+1,k+1)$ corresponds to the embedding of $M(n+k,k)$ in $M(n+k+1,k+1)$ given above, and so that the complement of $E(n+k,k)$ in $M(n+k+1,k+1)$ is $M(n+k,k+1)$ where $M(n+k,k+1) \hookrightarrow M(n+k+1,k+1)$ is given by $P \to P \oplus \{e_{n+k+1}\}$.

4. Show that $\begin{array}{c} E(n+k,k) \\ \downarrow \\ M(n+k,k) \end{array}$ can be identified with the normal bundle of the embedding of $M(n+k,k)$ in $M(n+k+1,k+1)$.

5. <u>Sheaf Theory</u>. Let $G$ be an abelian topological group, and let $X$ be a topological space. Let $\{U_i\}_{i \varepsilon J} = \mathcal{U}$ be an open covering of $X$. For $q \geq 0$, a $\mathcal{U}$ q-cochain is a function $f$ which associates to each

$(q+1)$ tuple $(i_o \cdots i_q)$, $i_j \in J$, a continuous map

$$f(i_o \cdots i_q): U_{i_o} \cap \cdots \cap U_{i_q} \to G.$$

The q-cochains $C^q(\mathcal{U}, S_G)$ form a group. Define $\delta^q: C^q(\mathcal{U}, S_G) \to C^{q+1}(\mathcal{U}, S_G)$ by

$$(\delta^q f)(i_o \cdots i_{q+1}) = \sum_{k=0}^{q+1} (-1)^k f(i_o \cdots \hat{i}_k \cdots i_{q+1}).$$

Then $\delta^{q+1}\delta^q = 0$.

Suppose $\mathcal{V} = \{V_\alpha\}_{\alpha\varepsilon\Lambda}$ is a refinement of $\mathcal{U}$. Pick a map $\tau: \Lambda \to J$ such that $V_\alpha \subseteq U_{\tau(\alpha)}$ for all $\alpha$ in $\Lambda$. Then the map $\tau^*: C^q(\mathcal{U}, S_G) \to C^q(\mathcal{V}: S_G)$ defined by $(\tau^* f)(\alpha_o \cdots \alpha_q) = f(\tau(\alpha_o), \ldots \tau(\alpha_q))$ is a chain map and the induced map $\tau^*: H^q(\mathcal{U}: S_G) \to H^q(\mathcal{V}: S_G)$ is independent of the choice of $\tau$. Define $H^q(X: S_G)$ to be the direct limit $\varinjlim_{\mathcal{U}} H^q(\mathcal{U}: S_G)$. Prove $H^1(X: S_G) \cong K_G(X)$, as abelian groups.

In the case, where $G$ is not abelian, we can still define $H^1(X: S_G)$ as a set with distinguished element.

Given an open covering $\mathcal{U} = \{U_i\}_{i\varepsilon J}$, a $\mathcal{U}$ cocycle $f$ associates to each pair $(i,j)$ of elements in $J$ a continuous map $f_{i,j}(U_i \cap U_j) \to G$ such that

$$f_{ij} \circ f_{jk} = f_{ik} \quad \text{on} \quad U_i \cap U_j \cap U_k.$$

Two co-cycles $f$ and $f'$ are equivalent if for each $j$ in $J$ there is a continuous map $g_j\colon U_j \to G$ such that

$$f'_{ij} = g_i^{-1} f_{ij} g_j \quad \text{on} \quad U_i \cap U_j.$$

Set $H^1(\mathcal{U}, S_G)$ = set of equivalence classes. Suppose $\mathcal{V} = \{V_\alpha\}_{\alpha\varepsilon\Lambda}$ is a refinement of $\mathcal{U} = \{U_j\}_{j\varepsilon J}$, pick a map $\tau\colon \Lambda \to J$ such that $V_\alpha \subset U_{\tau(\alpha)}$ for $\alpha$ in $\Lambda$. The map $\tau^*$ of $\mathcal{U}$ co-cycles into $\mathcal{V}$ co-cycles defined by requiring that

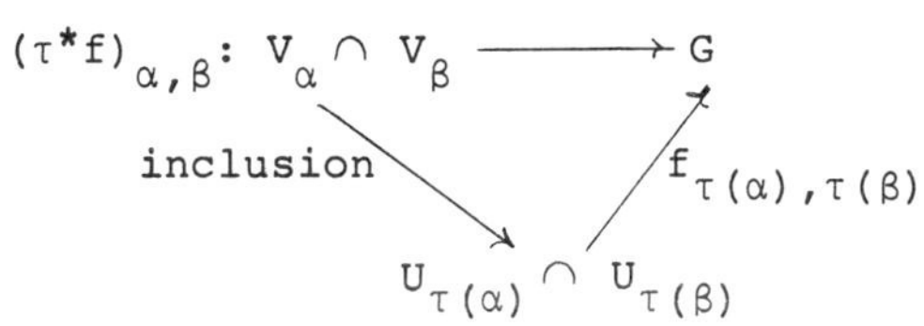


be commutative; induces a map $\tau^*\colon H^1(\mathcal{U}\colon S_G) \to H^1(\mathcal{V}\colon S_G)$ which is independent of $\tau$. Set $H^1(X\colon S_G) = \varinjlim_{\mathcal{U}} H^1(\mathcal{U}\colon S_G)$. Show $H^1(X\colon S_G) = K_G(X)$ as sets with distinguished element.

6. Show that a principle bundle is equivalent to a product bundle iff the bundle has a cross section. Thus a G-bundle is equivalent to a product bundle iff its associated principle bundle has a cross section.

7. Reduction of the group of a bundle to a closed subgroup.

   Let $H$ be a closed subgroup of $G$ and let

$h_*: K_H(B) \to K_G(B)$ be the map induced by the inclusion of $H$ into $G$. We say that the group of a coordinate bundle $\xi = (F,\tilde{E},B,\tilde{\pi},G,\{U_j\},\{\phi_j\})$ can be reduced to $H$ if the element $[\xi]$ in $K_G(B)$ is in the image of $H$.

(a) Show that the structure group of $\xi$ can be reduced to $H$ iff $\begin{array}{c}\tilde{E}\\ \downarrow \tilde{\pi}\\ B\end{array}$ can be given the structure of an $H$-bundle $\xi = (F,\tilde{E},B,\tilde{\pi},H,\{V_j\},\{\psi_j\})$ so that $(F,\tilde{E},B,\tilde{\pi},G,\{V_j\},\{\psi_j\})$ and $\xi$ are equivalent $G$-bundles. For the remainder of this problem, assume $H$ admits a local section in $G$.

(b) Let $(G,E,B,\pi,G,\{U_j\},\{g_j\})$ be a principal bundle and let $E/H$ be the identification space obtained by identifying, in each fibre of $E$, points which correspond under multiplication on the right by elements of $H$. Consider the commutative diagram

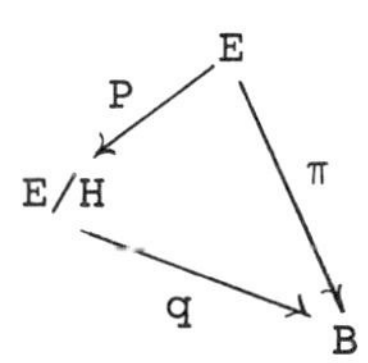


Show that $E$ can be given the structure of a principal bundle over $E/H$ with group $H$, and show that $E/H$ can be given the structure of a $G/H_o$-bundle over $B$ with

fibre $G/H$, and action induced by the map $G \times G/H \to G/H$ given by $(g,g'H) \to gg'H$. ($H_o$ = largest normal subgroup of $G$ which is a subgroup of $H$.)

(c) Let $\xi = (F,\tilde{E},B,\tilde{\pi},G,\{U_j\},\{\phi_j\})$ be a coordinate bundle. Let $(G,E,B,\pi,G,\{U_j\},\{g_j\})$ be the associated principal bundle. Set $\check{\xi} = (H,E,E/H,P,\{\ \},\{\ \})$, then

$$K_H(E/H) \xrightarrow{h} K_G(E/H) \xrightarrow{q^*} K_G(B)$$

$$[\check{\xi}] \longrightarrow h[\check{\xi}] = g^*[\xi] \longrightarrow [\xi].$$

In particular, the pullback of $E$ over $q$ is reducible.

(d) The structure group of $\xi$ can be reduced to $H$ iff the bundle $(G/H,E/H,B,q,\{\ \},\{\ \})$ has a cross section If a section $S$ is given, then the pullback of $E$ over $S$ is an H-bundle whose G-image is equivalent to $E$.

Chapter 3

# COVERING HOMOTOPY THEOREMS AND UNIVERSAL CLASSIFYING SPACES

## 1. Covering Homotopy Theorems

Theorem 3.1.1. (First Covering Homotopy Theorem). Let $(F,E,B,\pi,G,\{U_j\},\{\phi_j\})$ and $(F,E',B',\pi',G,\{V_j'\},\{\phi_j'\})$ be coordinate bundles, and let

$$\begin{array}{ccc} E & \xrightarrow{\tilde{H}_o} & E' \\ \downarrow{\scriptstyle\pi} & & \downarrow{\scriptstyle\pi'} \\ B & \xrightarrow{H_o} & B' \end{array}$$

be a map of coordinate bundles. Let $H\colon B \times I \to B'$ be a map with $H(b,0) = H_o(b)$ for all points b in B. Then, if B is a compact Hausdorff Space, there is a map

$$\begin{array}{ccc} E \times I & \xrightarrow{\tilde{H}} & E' \\ {\scriptstyle\pi\times 1_I}\downarrow & & \downarrow{\scriptstyle\pi'} \\ B \times I & \xrightarrow{H_o} & B' \end{array}$$

of coordinate bundles satisfying

1. $\tilde{H}$ restricted to $E \times \{0\}$ is the map $\tilde{H}_o$
2. $\tilde{H}$ is stationary with respect to the homotopy H. That is for any point e in E and any interval $[t_1,t_2] \subseteq I$ such that H is constant on $\{\pi(e)\} \times [t_1,t_2]$; $\tilde{H}$ must be constant on $\{e\} \times [t_1,t_2]$.

Recall: $E \times I \xrightarrow{\pi\times 1} B \times I$ is given the structure of a coordinate bundle $(F, E\times I, B\times I, \pi\times 1_I, G, \{U_j\times I\}, \{\phi_j\times 1\})$ by defining

$$\phi_j \times 1\colon (U_j \times I) \times F \to (\pi \times 1_I)^{-1}(U_j \times I)$$

to be the map

$$(u,t,x) \to [\phi_j(u,x),t].$$

The proof will make use of the following properties of Compact Hausdorff Spaces:

1. Given any point $b$ in a Compact Hausdorff Space $B$, and any nbhd $U$ of $b$, there is an open set $W$, whose closure, $\overline{W}$, is compact and such that $b \in W \subset \overline{W} \subset U$.

2. A Compact Hausdorff Space $B$, is normal and so we can use Urysohn's Lemma: i.e., Given disjoint closed subsets $K$ and $L$ of $B$, there is a continuous function $f\colon B \to [0,1]$ which is zero on $K$ and one on $L$.

Note: If $E' \xrightarrow{\pi'} B'$ is equivalent to the product bundle $B' \times F \xrightarrow{P_1} B'$, then $E$ must be equivalent to the product bundle $B \times F \xrightarrow{P_1} B$ and the theorem follows trivially.

Proof. The proof can be broken down into two main steps.

First Step. The first step is to show that there is an open covering $\{V_\ell\}_{\ell=1}^{K}$ of $B$ and a collection of numbers $\{a_i\}_{i=0}^{N}$, $0 = a_o < a_1 < a_2 < \cdots < a_{N-1} < a_N = 1$ such that each set

$V_\ell \times [a_i, a_{i+1}]$ is contained in $H^{-1}(V'_j)$ for some $V'_j$.

Second Step. The second step is to prove the theorem under the assumption that the open covering $\{H^{-1}(V'_j)\}$ of $B \times I$ has a refinement of the form $\{V_\ell \times I\}_{\ell=1}^{K}$ with $\{V\}_{\ell=1}^{K}$ an open covering of $B$.

Suppose the first and second steps have been proved. Using the first step we can pick an open cover $\{V\}_{\ell=1}^{K}$ of $B$ and a sequence $0 = a_o < a_1 < \cdots < a_{N-1} < a_N = 1$ such that each set $V_\ell \times [a_i, a_{i+1}]$ is contained in $H^{-1}(V'_j)$ for some $V'_j$. Using the second step extend $\tilde{H}_o$ to a map

$$\begin{array}{ccc} E \times [0,a_1] & \xrightarrow{\tilde{H}_{[0,a_1]}} & E' \\ {\scriptstyle \pi\times 1_{[0,a_1]}}\downarrow & & \downarrow{\scriptstyle \pi'} \\ B \times [0,a_1] & \xrightarrow{H} & B' \end{array}$$

of coordinate bundles so that $\tilde{H}_{[0,a_1]}$ is stationary with respect to $B \times [0,a_1] \xrightarrow{H} B'$. Now use the second step to extend the bundle map $E \times \{a_1\} \xrightarrow{\tilde{H}_{a_1}} E'$ to a map

$$\begin{array}{ccc} E \times [a_1,a_2] & \xrightarrow{\tilde{H}_{[a_1,a_2]}} & E' \\ {\scriptstyle \pi\times 1_{[a_1,a_2]}}\downarrow & & \downarrow{\scriptstyle \pi'} \\ B \times [a_1,a_2] & \xrightarrow{H} & B' \end{array}$$

of coordinate bundles so that $\tilde{H}_{[a_1,a_2]}$ is stationary with respect to $B \times [a_1,a_2] \xrightarrow{H} B'$.

Since $\tilde{H}_{[0,a_1]}$ and $\tilde{H}_{[a_1,a_2]}$ agree on $E \times \{a_1\}$, $\tilde{H}_{[0,a_1]}$ and $\tilde{H}_{[a_1,a_2]}$ define a continuous map $\tilde{H}_{[0,a_2]}: E \times [0,a_2] \to E'$. It follows that

$$\begin{array}{ccc} E \times [0,a_2] & \xrightarrow{\tilde{H}_{[0,a_2]}} & E' \\ \downarrow & & \downarrow \pi' \\ B \times I & \xrightarrow{H} & B' \end{array}$$

is a map of coordinate bundles and $\tilde{H}_{[0,a_2]}$ is stationary with respect to the map $B \times I \xrightarrow{H} B'$. Repeating this proceedure we can extend $\tilde{H}_{[0,a_2]}$ to a map $\tilde{H}: E \times I \to E'$ with the desired properties.

<u>Proof of the First Step</u>. Since sets of the form $\{V_\alpha \times I_\alpha : V_\alpha$ is open in $B, I_\alpha$ open in $I\}$ form a basis for the topology on $B \times I$, we can pick a refinement of $\{H^{-1}(V_j')\}$ of the form $\{V_\alpha \times I_\alpha\}_{\alpha \varepsilon \Lambda}$ with $V_\alpha$ open in $B$ and $I_\alpha$ open in $I$, $B \times I \bigcup_\alpha (V_\alpha \times I_\alpha)$. Since $B$ is compact, $B \times I$ is compact and so there is a finite subcover $\{V_i \times I_i\}_{i=1}^{s}$. For each $b$ in $B$ and $t$ in $I$ set $V_b = \bigcap_{b \varepsilon V_i} V_i$, $I_t = \bigcap_{t \varepsilon I_i} I_i$. The open cover $\{V_b \times I_t,\ b \varepsilon B,\ t \varepsilon I\}$

is a refinement of $\{V_i \times I_i\}_{i=1}^{s}$ and hence a refinement of $\{H^{-1}(V_j')\}$. Pick $V_{b_1} \cdots V_{b_\ell}$, $I_{t_1} \cdots I_{t_r}$ such that $B \subset \bigcup_{i=1}^{\ell} V_{b_i}$, $I \subset \bigcup_{k=1}^{r} I_{t_k}$.

Then $\left\{ V_{b_i} \times I_{t_k} \right\}_{\substack{i=1\cdots\ell \\ k=1\cdots r}}$ is an open cover of $B \times I$ and a refinement of $\{H^{-1}(V_j')\}$. It is not hard to see that any open cover of $I$ has a refinement, $\{I_j\}_{j=1}^{N}$ satisfying

(a) $I_j$ is the intersection of an open interval with $I$.

(b) $I_j \cap I_k = \emptyset$ unless $k = j - 1$, $j$ or $j + 1$.

Thus, there is a refinement $\{I_j\}_{j=1}^{N}$ of $\left\{ I_{t_k} \right\}_{k=1}^{r}$ which covers $I$ and satisfies (a) and (b) above. Given such a refinement $\{I_j\}_{j=1}^{N}$, pick points $0 = a_o < a_1 < a_2 \cdots\cdots < a_{N-1} < a_r = 1$ with $a_j \in I_j \cap I_{j+1}$ for $j = 1 \cdots N - 1$.

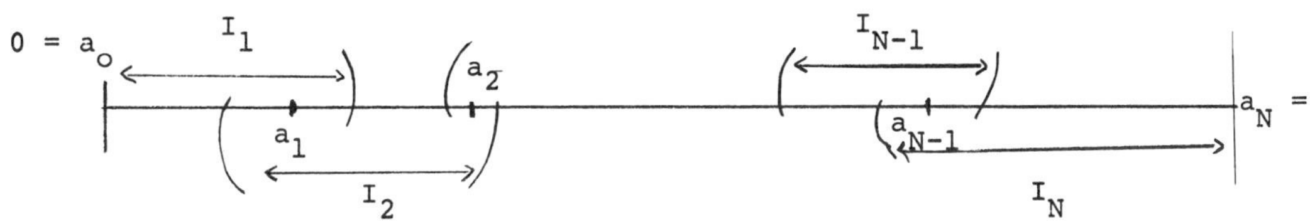


Diagram 7.

Then $\left\{ V_{b_i} \times I_j \right\}_{\substack{i=1\cdots\ell \\ j=1\cdots N}}$ is a refinement of $H^{-1}(V_j')$ so each set

$V_{b_i} \times [a_j, a_{j+1}]$ is contained in $H^{-1}(V_j')$ for some $V_j'$ as desired. This completes the proof of the first step.

<u>Proof of the Second Step</u>. Assume the open covering $\{H^{-1}(V_j')\}$ of $B \times I$ has a refinement of the form $\{V_\ell \times I\}_{\ell=1}^k$ with $\{V_\ell\}_{\ell=1}^k$ an open covering of $B$. For each point $b$ in $B$ choose a pair of open sets $(W, W')$ such that $b \in W \subset \overline{W} \subset W' \subset \overline{W}'$ with $\overline{W}'$ contained in some $V_\ell$. Pick a finite collection of such pairs $(W_1, W_1') \cdots (W_s, W_s')$ so that $B \subset \bigcup_{\alpha=1}^{s} W_\alpha$. From Urysohn's Lemma, it follows that for each $\alpha = 1 \cdots s$ there is a map $u_\alpha \colon B \to [0,1]$ such that $u_\alpha(\overline{W}_\alpha) = 1$ and $u_\alpha(B - W_\alpha') = 0$. Define $t_0 \colon B \to [0,1]$ by $t_0(b) = 0$ for all $b$ in $B$. For $1 \leq \alpha \leq s$ set $t_\alpha(b) = \max\{u_1(b), \ldots u_\alpha(b)\}$ then

1. each $t_\alpha \colon B \to [0,1]$ is continuous
2. $0 = t_0(b) \leq t_1(b) \leq t_2(b) \leq \cdots \leq t_s(b) = 1$ for all $b$ in $B$

and

3. for $\alpha = 1 \cdots s$, $\{b \text{ in } B\colon t_{\alpha-1}(b) < t_\alpha(b)\} \subset W_\alpha'$.

Set

$$(B \times I)_\alpha = \{(b,t)\colon 0 \leq t \leq t_\alpha(b)\} \quad \alpha = 0, \ldots s$$

and

$$(E \times I)_\alpha = (\pi \times 1)^{-1}(B \times I)_\alpha = \{(e,t): 0 \leq t \leq t_\alpha(e)\}.$$

Note: $(E \times I)_o = E \times \{0\}$, $(B \times I)_o = B \times \{0\}$.

$$(B\times I)_o \subset (B\times I)_1 \subset \cdots \subset (B\times I)_\alpha \subset (B\times I)_{\alpha+1} \subset \cdots \subset (B\times I)_s = B \times I$$

$$(E\times I)_o \subset (E\times I)_1 \subset \cdots \subset (E\times I)_\alpha \subset (E\times I)_{\alpha+1} \subset \cdots \subset (E\times I)_s = E \times I$$

Each $(B \times I)_\alpha$ is closed and the closure of $(B \times I)_{\alpha+1} - (B \times I)_\alpha$ is contained in $\{(b,t): t_\alpha(b)\} \leq t \leq t_{\alpha+1}(b)\} \cap (\overline{W}'_{\alpha+1} \times I)$.

For each $\alpha$, $\begin{array}{c}(E \times I)_\alpha \\ \downarrow{\scriptstyle \pi\times 1} \\ (B \times I)_\alpha\end{array}$ is a coordinate bundle. To complete the proof of the second step it suffices to show that if $\tilde{H}_o$ has been extended to a stationary bundle map

$$\begin{array}{ccc}(E \times I)_\alpha & \xrightarrow{\tilde{H}_\alpha} & E' \\ \downarrow{\scriptstyle (\pi\times 1)} & & \downarrow{\scriptstyle \pi'} \\ (B \times I)_\alpha & \xrightarrow{H} & B'\end{array}$$

then $\tilde{H}_\alpha$ can be extended to $(E \times I)_{\alpha+1}$. For each $\alpha \geq 0$ define $f_{\alpha+1}: (B \times I)_{\alpha+1} \to (B \times I)_\alpha$ by

$$f_{\alpha+1}(b,t) = \left\{ \begin{array}{ll} (b,t) & \text{if} \quad 0 \leq t \leq t_\alpha(b) \\ (b,t_\alpha(b)) & \text{if} \quad t_\alpha(b) \leq t \leq t_{\alpha+1}(b) \end{array} \right\}$$

$f_{\alpha+1}(b,t)$ is continuous since it is continuous when restricted to each of the closed sets $(B \times I)_\alpha$ and $\{(b,t): t_\alpha(b) \leq t \leq t_{\alpha+1}(b)\}$. Similarly define $\tilde{f}_{\alpha+1}(E \times I)_{\alpha+1} \to (E \times I)$ by

$$\tilde{f}_{\alpha+1}(e,t) = \left\{ \begin{array}{l} (e,t) \quad \text{if} \quad 0 \leq t \leq t_\alpha \pi(e) \\ (e,t_\alpha \pi(e)) \quad \text{if} \\ \qquad t_\alpha \pi(e) \leq t \leq t_{\alpha+1}\pi(e) \end{array} \right\} .$$

Assume

$$\begin{array}{ccc} (E \times I)_\alpha & \xrightarrow{\tilde{H}_\alpha} & E' \\ \downarrow & & \downarrow \pi' \\ (B \times I)_\alpha & \xrightarrow{H} & B' \end{array}$$

is a stationary map of coordinate bundles.

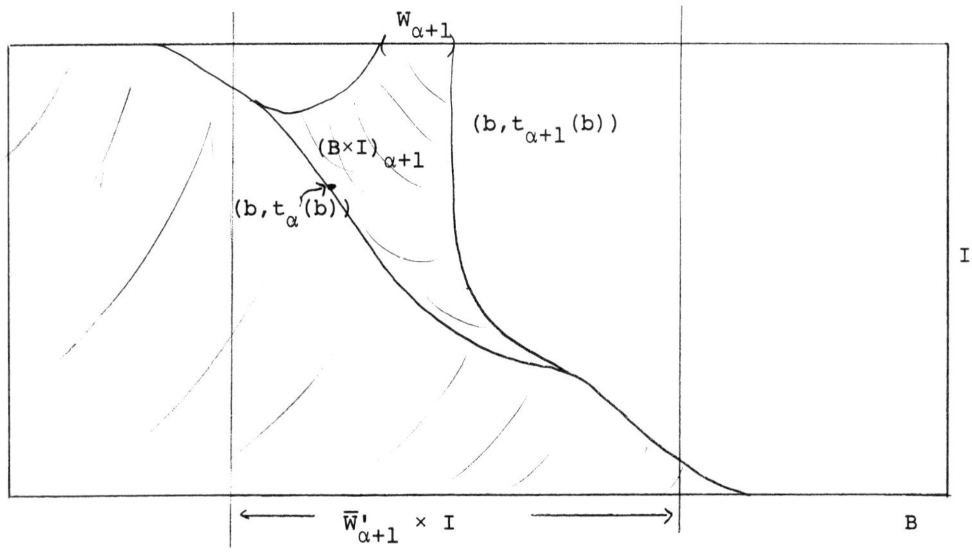


Diagram 8.

$$(B\times I)_{\alpha+1} - (B\times I)_{\alpha} \subset \{(b,t): t_\alpha(b) \leq t \leq t_{\alpha+1}(b)\} \cap (\overline{W}'_{\alpha+1}\times I)$$

$$\subset (B\times I)_{\alpha+1}.$$

Since $\{\overline{W}'_\alpha \times I\}_{\alpha=1}^{s}$ is a refinement of $\{H^{-1}(V'_j)\}$, we can pick a $V'_j$ such that $\{(b,t): t_\alpha(b) \leq t \leq t_{\alpha+1}(b)\} \cap (\overline{W}'_{\alpha+1} \times I) \subset H^{-1}(V'_j)$. Set $\tilde{H}_{\alpha+1} = \tilde{H}_\alpha$ on $(E \times I)_\alpha$ and define $\tilde{H}_{\alpha+1}$ on $(\pi \times 1)^{-1}(\{(b,t): t_\alpha(b) \leq t \leq t_{\alpha+1}(b)\} \cap (\overline{W}'_{\alpha+1} \times I))$ by $(e,t) \to \phi'_j(H(\pi(e),t)), P_2\phi'^{-1}_j\tilde{H}_\alpha\tilde{f}_{\alpha+1}(e,t))$.

$\tilde{H}_{\alpha+1}$: $(E \times I)_{\alpha+1} \to E'$ is well defined and continuous. From the definition of $\tilde{f}_{\alpha+1}$ it follows that if $\tilde{H}_\alpha$ is stationary so is $\tilde{H}_{\alpha+1}$. It remains to show that $\tilde{H}_{\alpha+1}$ is a map of coordinate bundles. That is for each nonempty intersection of the form

$[(U_i \times I) \cap H^{-1}(V_k')] \cap (B \times I)_{\alpha+1}$ there is a continuous map $(g_{k',i})_{\alpha+1}([(U_i \times I) \cap H^{-1}(V_k')] \cap (B \times I)_{\alpha+1}) \to G$ so that the composition

$$\begin{array}{ccc}
([(U_i \times I) \cap H^{-1}(V_k')] \cap (B \times I)_{\alpha+1}) \times F & & V_k' \times F \\
\Big\downarrow \phi_i \times 1 & & \Big\uparrow \phi_{k'}^{-1} \\
(\pi \times 1)^{-1}([(U_i \times I) \cap H^{-1}(V_k')] \cap (B \times I)_{\alpha+1}) & \xrightarrow{\tilde{H}_{\alpha+1}} & (\pi')^{-1}(V_k')
\end{array}$$

is given by $((b,t),x) \to (H(b,t),(g_{k',i})_{\alpha+1}(b,t) \cdot x)$. Assume $[(U_i \times I) \cap H^{-1}(V_k')] \cap (B \times I)_{\alpha+1} \neq \emptyset$. Since $\{V_\ell \times I\}_{\ell=1}^{k}$ is a refinement of $H^{-1}(V_k')$, it follows that $[(U_i \times I) \cap H^{-1}(V_k')] \cap (B \times I)_\alpha \neq \emptyset$ and that $f_{\alpha+1}$ induces a continuous map

$$[(U_i \times I) \cap H^{-1}(V_k')] \cap (B \times I)_{\alpha+1} \xrightarrow{f_{\alpha+1}} [(U_i \times I) \cap H^{-1}(V_k')] \cap (B \times I)_\alpha$$

Define $(g_{k',i})_{\alpha+1}$ by requiring that

$$\begin{array}{ccc}
[(U_i \times I) \cap H^{-1}(V_k')] \cap (B \times I)_{\alpha+1} & \xrightarrow{(g_{k',i}')_{\alpha+1}} & G \\
& \searrow f_{\alpha+1} \qquad \nearrow (g_{k',i})_\alpha & \\
& [(U_i \times I) \cap H^{-1}(V_k')] \cap (B \times I)_\alpha &
\end{array}$$

be commutative. Our assumption that $\tilde{H}_\alpha$ be a bundle map implies the existence of the map $(g_{k',i})_\alpha$. The definitions of $\tilde{f}_{\alpha+1}, f_{\alpha+1}$ and $\tilde{H}_{\alpha+1}$ guarantee that $(g_{k',i})_{\alpha+1}$ has the desired property and $\tilde{H}_{\alpha+1}$ is a map of coordinate bundles. This completes the proof of the first covering homotopy theorem.

<u>Theorem 3.1.2</u>. If B is a compact Hausdorff space, then any bundle over $B \times I$ is equivalent to a product bundle of the form $E \times I$.

<u>Proof</u>. Let $\begin{array}{c}\tilde{E}\\ \downarrow\\ B \times I\end{array}$ be a coordinate bundle. Let E be the pullback of $\tilde{E}$ over the map $u_o: B \to B \times I$ where $u_o(b) = (b,0)$ for all b in B. Applying the first covering homotopy theorem, we get a map of coordinate bundles

$$\begin{array}{ccc} E \times I & \xrightarrow{\ \tilde{H}\ } & \tilde{E} \\ \downarrow & & \downarrow \\ B \times I & \xrightarrow{\text{identity}} & B \times I \end{array}$$

Since $\tilde{H}$ covers the identity map of $B \times I$ into itself, $\tilde{H}$ is an equivalence of coordinate bundles.

<u>Theorem 3.1.3</u>. Let $\begin{array}{c}E'\\ \downarrow \pi'\\ B'\end{array}$ be a coordinate bundle, let $f_o$ and $f_1$ be homotopic maps of the compact Hausdorff space B into E'. Then the pullbacks $E_{f_o}$ and $E_{f_1}$ are equivalent coordinate bundles.

Proof. Let $H\colon B \times I \to B'$ be such that $H(b,0) = f_o(b)$ and $H(b,1) = f_1(b)$ for all $b$ in $B$.

From the first covering homotopy theorem it follows that there is a map of coordinate bundles

$$\begin{array}{ccc} E_{f_o} \times I & \xrightarrow{\tilde{H}} & E' \\ \downarrow & & \downarrow \pi' \\ B \times I & \xrightarrow{H} & B' \end{array}$$

Define $u_1\colon B \to B \times I$ by $u_1(b) = (b,1)$ and $E_{f_o} \xrightarrow{\tilde{u}_1} E_{f_o} \times I$ by $\tilde{u}_1(e) = (e,1)$. Then

$$\begin{array}{ccc} E_{f_o} & \xrightarrow{\tilde{u}_1} & E_{f_o} \times I \\ \downarrow & & \downarrow \\ B & \xrightarrow{u_1} & B \times I \end{array}$$

is a map of coordinate bundles. The composition of maps of coordinate bundles, is a map of coordinate bundles. Thus

$$\begin{array}{ccc} E_{f_o} & \xrightarrow{\tilde{H}o\tilde{u}_1} & E' \\ \downarrow & & \downarrow \pi' \\ B & \xrightarrow{f_1 = Hou_1} & B' \end{array}$$

is a map of coordinate bundles

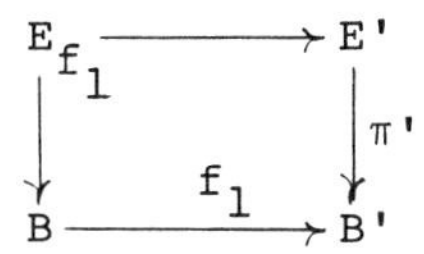


is also a map of coordinate bundles and it follows that $E_{f_o}$ and $E_{f_1}$ are equivalent coordinate bundles.

<u>Corollary 3.1.4</u>. If $B \xrightarrow{f} B'$ is a homotopy equivalence of compact Hausdorff spaces, then $f^* K_G(B') \to K_G(B)$ is 1-1 and onto.

<u>Proof</u>. Let $g\colon B' \to B$ be a homotopy inverse of $f$, i.e., $f \circ g$ is homotopic to the identity map of $B'$ onto itself and $g \circ f$ is homotopic to the identity map of $B$ onto itself. From the previous theorem it follows that

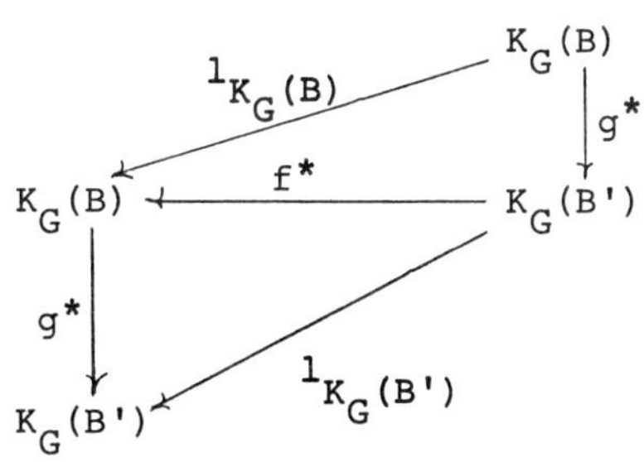


commutes and the result follows.

<u>Theorem 3.1.5.</u> Second Covering Homotopy Theorem. Let $B$ be a compact Hausdorff space, $\begin{array}{c} E' \\ \downarrow \pi' \\ B' \end{array}$ a coordinate bundle, 

$$\begin{array}{ccc} & & E' \\ & \overset{\tilde{f}_o}{\nearrow} & \downarrow \pi' \\ B & \xrightarrow{f_o} & B' \end{array}$$

commutative, and $H\colon B \times I \to B'$ a continuous map with $H(b,0) = f_o(b)$ for all $b$ in $B$. Then there is a map $\tilde{f}\colon B \times I \to E'$ such that

$$\begin{array}{ccc} & & E' \\ & \overset{\tilde{f}}{\nearrow} & \downarrow \pi' \\ B \times I & \xrightarrow{H} & B' \end{array}$$

commutes, $\tilde{f}(b,0) = \tilde{f}_o(b)$ for all $b$ in $B$ and $\tilde{f}$ is stationary with respect to $H$.

<u>Proof.</u> Let $E_{f_o}$ be the pullback of $E'$ over $f_o$. Then there is a map of coordinate bundles

$$\begin{array}{ccc} E_{f_o} & \xrightarrow{\tilde{H}_o} & E' \\ \downarrow & & \downarrow \pi' \\ B & \xrightarrow{f_o} & B' \end{array}$$

and from the first covering homotopy theorem it follows that there is a map of coordinate bundles

$$\begin{array}{ccc} E_{f_o} \times I & \xrightarrow{\tilde{H}} & E' \\ \downarrow & & \downarrow \pi' \\ B \times I & \xrightarrow{H} & B' \end{array}$$

Recall that $E_{f_o}$ is, by definition, the subspace of $B \times E'$ consisting of pairs $(b,e')$ such that $f_o(b) = \pi'(e')$. Define $c\colon B \to E_{f_o}$ by $c(b) = (b,f_o(b)$ and $(c \times 1)\colon B \times I \to E_{f_o} \times I$ by $(c \times 1)(b,t) = (c(b),t)$. Define $\tilde{f}$ to be the composition $\tilde{H}\circ(c \times 1)$.

The second covering homotopy theorem shows that for compact Hausdorff spaces $B$, coordinate bundles $\begin{array}{c} E' \\ \downarrow \pi' \\ B' \end{array}$ and maps $f\colon B \to B'$; the existence or nonexistence of a lifting of $f$ to $E'$ is determined by the homotopy class of $f$.

## 2. A Quick Review of Homotopy Theory, the Exact Homotopy Sequence of a Coordinate Bundle

Let $X$ be a topological space, $A$ a subspace of $X$ and $x_o$ a point in $A$. For each $n \geq 1$ set $I^n = \{(t_1,t_2,\ldots t_n)\colon 0 \leq t_i \leq 1\}$ and $\dot{I}^n = \{(t_1\cdots t_n)\colon t_i = 0 \text{ or } 1 \text{ for some } i\}$.

For $n > 1$ set $I^{n-1} = \{(t_1\cdots t_n) \text{ in } I^n\colon t_n = 0\}$ and $J^{n-1} = \{(t_1\cdots t_n) \text{ in } I^n\colon t_i = 0 \text{ for some } i \text{ with } 1 \leq i \leq n-1\}$. Then $\dot{I}^n = I^{n-1} \cup J^{n-1}$.

For $n \geq 2$ define the relation $\sim$ on the set of maps $f\colon (I^n,I^{n-1},J^{n-1}) \to (X,A,x_o)$ by setting $f \sim g$ if there is a map

$$H\colon (I^n \times I,\ I^{n-1} \times I,\ J^{n-1} \times I) \to (X,A,x_o)$$

such that $H$ restricted to $I \times \{0\}$ is $f$ and $H$ restricted to

$I \times \{1\}$ is $g$. $\sim$ is an equivalence relation.

Set $\pi_n(X,A,x_o)$ equal to the set of equivalence classes and denote by $[f]$ the element of $\pi_n(X,A,x_o)$ determined by the map $f\colon (J^n,I^{n-1},J^{n-1}) \to (X,A,x_o)$.

Given maps $f\colon (I^n,I^{n-1},J^{n-1}) \to (X,A,x_o)$ and $g\colon (I^n,I^{n-1},J^{n-1}) \to (X,A,x_o)$. Define $(f+g)\colon (I^n,I^{n-1},J^{n-1}) \to (X,A,x_o)$ by

$$(f+g)(t_1\cdots t_n) = \begin{cases} f(2t_1,t_2\cdots t_n) & 0 \le t_1 \le \frac{1}{2} \\ g(2t_1-1,\ t_2\cdots t_n) & \frac{1}{2} \le t_1 \le 1 \end{cases}.$$

If $[f] = [f']$ and $[g] = [g']$ then $[f+g] = [f'+g']$. Thus $(f,g) \to f + g$ induces a map

$$\pi_n(X,A,x_o) \times \pi_n(X,A,x_o) \to \pi_n(X,A,x_o)$$

$$[f],[g] \to [f] + [g] = [f+g].$$

It can be shown that $([f] + [g]) + [h] = [f] + ([g] + [h])$.

Let $f\colon (I^n,I^{n-1},J^{n-1}) \to (X,A,x_o)$ be the constant map $f(t_1,\ldots t_n) = x_o$ for all $(t_1\cdots t_n)$ in I, and let $e = [f]$. Then $e + [g] = [g] + e = [g]$ for all $[g]$ in $\pi_n(X,A,x_o)$.

Given a map $f\colon (I^n,I^{n-1},J^{n-1}) \to (X,A,x_o)$, set

$$\bar{f}(t_1\cdots t_n) = f(1-t_1,t_2,\ldots t_n).$$

Note

$$f + \bar{f}(t_1 \cdots t_n) = \left\{ \begin{array}{ll} f(2t_1, t_2 \cdots t_n) & 0 \le t_1 \le \frac{1}{2} \\ f(2(1-t_1),\ t_2 \cdots t_n) & \frac{1}{2} \le t_1 \le 1 \end{array} \right\} .$$

Define $F: (I^n \times I,\ I^{n-1} \times I,\ J^{n-1} \times I) \to (X, A, x_o)$ by

$$F(t_1 \cdots t_n, s) = \left\{ \begin{array}{ll} f(2t_1, t_2 \cdots t_n) & 0 \le t_1 \le \frac{1}{2} s \\ f(s, t_2 \cdots t_n) & \frac{1}{2} s \le t_1 \le 1 - \frac{1}{2} s \\ f(2(1-t_1),\ t_2 \cdots t_n) & 1 - \frac{1}{2} s \le t_1 \le 1 \end{array} \right\} .$$

Then

$$F(t_1 \cdots t_n, 0) = f(0, t_2 \cdots t_n) = x_o$$

and

$$F(t_1 \cdots t_n, 1) = (f+\bar{f})(t_1 \cdots t_n).$$

Thus $[f] + [\bar{f}] = e$. Since $\bar{\bar{f}} = f$, $[\bar{f}]$ is a two sided inverse of $[f]$. This shows that $\pi_n(X,A,x_o)$ is a group for $n \ge 2$. $+$ is used to denote the composition because $\pi_n(X,A,x_o)$ is abelian for $n > 2$.

To see that $\pi_n(X,A,x_o)$ is abelian for $n > 2$, let $h$ be a homeomorphism of $I^2$ onto $E^2 = \{(x_1,x_2): x_1^2 + x_2^2 \le 1\}$ such that

$$\begin{array}{ccc} I^2 & \longrightarrow & I^2 \\ h\big\downarrow & & \big\downarrow h \\ E^2 & \longrightarrow & E^2 \end{array}$$

commutes where $I^2 \to I^2$ is given by $(t_1,t_2) \to (1-t_1,1-t_2)$ and $E^2 \to E^2$ is defined by $(x_1,x_2) \to (-x_1,-x_2)$. For $n \geq 2$ define $H\colon I^n \times I \to I^n$ by $H(t_1,t_2,\ldots t_n,s) = (h^{-1}e^{i\pi s}h(t_1,t_2),\ t_3\cdots t_n)$. Then

$$H(t_1,t_2,\ldots t_n,0) = (t_1,t_2,\ldots t_n)$$

and

$$H(t_1,t_2,\ldots t_n,1) = (1-t_1,1-t_2,t_3\cdots t_n).$$

For $n > 2$, $H$ is a map of triples $(I^n \times I,\ I^{n-1} \times I,\ J^{n-1} \times I) \to (I^n,I^{n-1},J^{n-1})$. Given $f\colon (I^n,I^{n-1},J^{n-1}) \to (X,A,x_o)$ define $f^\pi$ by $f^\pi(t_1,t_2,\ldots t_n) = f(1-t_1,1-t_2,t_3\cdots t_n)$. The above shows that for $n > 2$ $[f] = [f^\pi]$ and

$$[f] + [g] = [f+g] = [(f+g)^\pi] = [g^\pi+f^\pi] = [g^\pi] + [f^\pi] = [g] + [f].$$

So $\pi_n(X,A,x_o)$ is abelian for $n > 2$.

Given $X$ and a point $x_o$ in $X$. Define the relation $\sim$ on maps $f\colon (I^n,\dot{I}^n) \to (X,x_o)$ for $n \geq 1$ by $f \sim g$ if there is a map $H\colon (I^n \times I,\ \dot{I}^n \times I) \to (X,x_o)$ such that $H$ restricted to $I^n \times \{1\}$ is $f$ and $H$ restricted to $I^n \times \{0\}$ is $g$.

$\sim$ is an equivalence relation. Set $\pi_n(X,x_o)$ equal to the set of equivalence classes. Then $\pi_n(X,x_o)$ is a group for $n \geq 1$, $\pi_n(X,x_o)$ is abelian for $n \geq 2$, and $\pi_n(X,x_o)$ is isomorphic to $\pi_n(X,x_o,x_o)$ for $n \geq 2$.

A map $f(X,A,x_o) \to (Y,B,y_o)$ induces a homomorphism of groups $f_*: \pi_n(X,A,x_o) \to \pi_n(Y,B,y_o)$ for $n \geq 2$, and a map $g: (X,x_o) \to (Y,y_o)$ induces a homomorphism $\pi_n(X,x_o) \to \pi_n(Y,y_o)$ for $n \geq 1$. Given $f: (I^n,I^{n-1},J^{n-1}) \to (X,A,x_o)$ define $\partial f: (I^{n-1},\dot{I}^{n-1}) \to (A,x_o)$ to be the restriction of $f$ to $I^{n-1}$. Then $f \to \partial f$ induces a homomorphism $\partial: \pi_n(X,A,x_o) \to \pi_{n-1}(A,x_o)$ for each $n \geq 2$.

Let $i: (A,x_o) \to (X,x_o)$ and $j: (X,x_o,x_o) \to (X,A,x_o)$ be the inclusion maps, then

$$\cdots \to \pi_n(A,x_o) \xrightarrow{i_*} \pi_n(X,x_o) \xrightarrow{j_*} \pi_n(X,A,x_o) \xrightarrow{\partial} \pi_{n-1}(A,x_o) \to \cdots$$

$$\cdots \to \pi_n(X,A,x_o) \xrightarrow{\partial} \pi_1(A,x_o) \xrightarrow{i_*} \pi_1(X,x_o)$$

is exact.

For $n \geq 2$, there is a homeomorphism $(I^n/J^{n-1},\ I^{n-1}/I^{n-1} \cap J^{n-1},\ J^{n-1}/J^{n-1}) \to (E^n,S^{n-1},*)$ where $I^n/J^{n-1}$ denotes the quotient space of $I^n$ obtained by identifying all the points in $J^{n-1}$ to a single point, $E^n = \{(x_1 \cdots x_n): \Sigma\, x_i^2 \leq 1\}$, $S^{n-1} = \{(x_1 \cdots x_n): \Sigma\, x_i^2 = 1\}$ and $*$ is the point $(1,0,\ldots 0)$. The corresponding quotient map $(I^n,I^{n-1},J^{n-1}) \to (E^n,S^{n-1},*)$ sets up a 1-1 correspondence between maps of $(I^n,I^{n-1},J^{n-1})$ into $(X,A,x_o)$ and maps of $(E^n,S^{n-1},*)$ into $(X,A,x_o)$

Similarly there is for $n \geq 1$, a homeomorphism $(I^n/\dot{I}^n, \dot{I}^n/\dot{I}^n) \to (S^n,*)$ which sets up a 1-1 correspondence between maps of $(I^n,\dot{I}^n)$ into $(X,x_o)$ and maps of $(S^n,*)$ into $(X,x_o)$. Thus we can view $\pi_n(X,A,x_o)$ as equivalence classes of maps of $(E^n,S^{n-1},*)$ into $(X,A,x_o)$ with equivalence relation defined by $f \sim g$ iff there is a map $H: (E^n \times I, S^{n-1} \times I, * \times I) \to (X,A,x_o)$ such that H-restricted to $E^n \times 0$ is $f$ and $H$ restricted to $E^n \times 1$ is $g$.

Similarly $\pi_n(X,x_o)$ can be viewed as equivalence classes of maps $(S^n,*) \to (X,x_o)$ with equivalence relation defined by $f \sim g$ iff there is a map $H: (S^n \times I, * \times I) \to (X,x_o)$ such that $H$ restricted to $S^n \times 0$ is $f$ and $H$ restricted to $S^n \times 1$ is $g$. Define $\sigma: S^{n-1} \times I \to E^n$ by $(x,t) \to xt + *(1-t)$. Suppose a map $f: (S^{n-1},*) \to (X,x_o)$ represents the zero element of $\pi_{n-1}(X,x_o)$, then there is a map $H(S^{n-1} \times I, * \times I) \to (X,x_o)$ such that $H$ restricted to $S^{n-1} \times 1$ is $f$ and $H$ restricted to $S^{n-1} \times 0$ is the constant map $x_o$. It follows that $H$ factors into the composition of a map $H'$ of $E^n$ to $X$ and $\sigma$

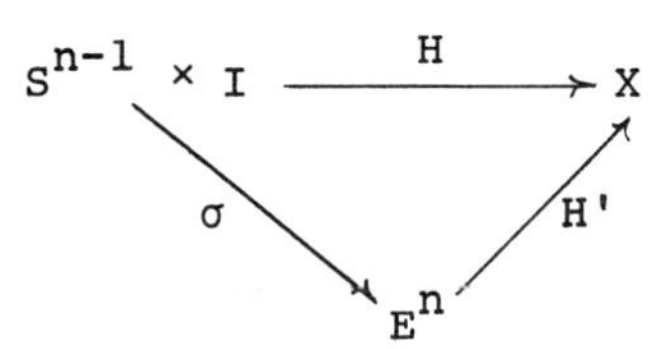


so that $H'$ restricted to $S^{n-1} \subseteq E^n$ is the map $f$. Conversely if the map $f: (S^{n-1},*) \to (X,x_o)$ can be extended to a map $H'$ of $E^n$ into $X$, then the composition

$$S^{n-1} \times I \xrightarrow{\sigma} E^n \xrightarrow{H'} X$$

is a homotopy of $f$ to the constant map, showing that $[f] = 0$ in $\pi_{n-1}(X,x_o)$. Thus $[f] = 0$ in $\pi_{n-1}(X,x_o)$ iff $f\colon (S^{n-1},*) \to (X,x_o)$ can be extended to a map of $E^n$ into $X$.

Using the map $\bar{\sigma}\colon E^n \times I \to E^n$ defined by $(x,t) \to xt + *(1-t)$ it can be seen that a map $f\colon (E^n,S^{n-1},*) \to (X,A,x_o)$ iff there is a map $H\colon (E^n \times I, S^{n-1} \times I, * \times I) \to (X,A,x_o)$ such that $H$ restricted to $E^n \times 0$ is $f$ and $H$ restricted to $E^n \times 1$ is a map of $E^n$ into $A$. That is using $\bar{\sigma}$, we see that any map $f'(E^n,S^{n-1},*) \to (X,A,x_o)$ with $f'(E^n) \subseteq A$ represents the zero element of $\pi_n(X,A,x_o)$.

In [Whitehead], it is proven that if $X$ and $Y$ are C. W. complexes, $L$ is a subcomplex of $X$, and $f$ is a map of $X$ into $Y$ such that $f$ restricted to $L$ is cellular; then there is a homotopy $H\colon X \times I \to Y$ such that $H$ restricted to $X \times 1$ is $f$, $H(\ell,t) = f(\ell)$ for $t \in I$, and $H$ restricted to $X \times 0$ is cellular. This result will be referred to as Whitehead's Cellular Approximation Theorem. It follows that if $X$ is a C. W. complex with 0-cell $x_o$ and $X^n = X^o$, then $\pi_i(X,x_o) = 0$ for $1 \leq i \leq n$. In particular, $\pi_i(G(n,k),\{I\}) = 0$ for $1 \leq i \leq (k+1)d-2$ and $\pi_i(S^n,*) = 0$ for $1 \leq i < n$.

Recall the definition of cubical singular homology theory. Set $I^o = pt$, and let $X$ be a topological space. For $n \geq 0$ set $Q_n(X)$ equal to the free abelian group generated by all continuous maps $f\colon I^n \to X$. A map $f\colon I^n \to X$ is called degenerate if there is an integer $i$, $1 \leq i \leq n$, such that $f(t_1 \cdots t_n)$ does not depend on $t_i$.

Let $D_n(X)$ be the subgroup of $Q_n(X)$ generated by the degenerate maps $f: I^n \to X$. For $n \geq 0$ set $C_n(X) = Q_n(X)/D_n(X)$. $C_n(X)$ is called the group of cubical singular n-chains of X. The faces of a map $T: I^n \to X$ for $n > 0$ are defined by

$$A_iT(t_1 \cdots t_{n-1}) = T(t_1 \cdots t_{i-1}, 0, t_i, \ldots t_{n-1})$$

$$B_iT(t_1 \cdots t_{n-1}) = T(t_1 \cdots t_{i-1}, 1, t_i \cdots t_{n-1}).$$

For $T: I^n \to X$, $n > 1$, $1 \leq i < j \leq n$ we have the following identities

$$A_iA_j(T) = A_{j-1}A_i(T)$$

$$B_iB_j(T) = B_{j-1}B_i(T)$$

$$A_iB_j(T) = B_{j-1}A_i(T)$$

$$B_iA_j(T) = A_{j-1}B_i(T)$$

Define $\partial_n: Q_n(X) \to Q_{n-1}(X)$ for $n \geq 1$ by $\partial_n(T) = \sum_{i=1}^{n} (-1)^i [A_iT - B_iT]$.

Then $\partial_{n-1}\partial_n \equiv 0$ for $n > 1$ and $\partial_n(D_n(X)) \subset D_{n-1}(X)$ for $n > 0$. Thus $\partial_n$ induces a map $C_n(X) \xrightarrow{\partial_n} C_{n-1}(X)$. The homology groups, $H_i(C_*(X), \partial)$, denoted by $H_i(X)$, are called cubical singular homology groups of X. If A is a subspace of X, then

$$0 \longrightarrow C_*(A) \xrightarrow{i} C_*(X) \xrightarrow{j} C_*(X)/C_*(A) \longrightarrow 0$$

is a short exact sequence of chain complexes. The corresponding long exact sequence is denoted by

$$\cdots \to H_n(A) \xrightarrow{i_*} H_n(X) \xrightarrow{j_*} H_n(X,A) \xrightarrow{\partial} H_{n-1}(A) \xrightarrow{i_*} H_{n-1}(X) \longrightarrow \cdots$$

$$\cdots \to H_1(A) \xrightarrow{i_*} H_1(X) \xrightarrow{j_*} H_1(X,A) \xrightarrow{\partial} H_o(A) \xrightarrow{i_*} H_o(X) \xrightarrow{j_*} H_o(X,A) \to 0.$$

Let $f$ be a map $f\colon (I^n, I^{n-1}, J^{n-1}) \to (X,A,x_o)$. $f\colon I^n \to X$ is a singular n-cube and since $f$ maps $\dot{I}^n$ into $A$, the projection of $f$ to $C_n(X)/C_n(A)$ is a cycle. Set $\phi(f)$ equal to the homology class of the cycle $j(f)$. For $n \geq 2$, $\phi$ induces a homomorphism

$$\phi\colon \pi_n(X,A,x_o) \to H_n(X,A) \quad \text{for} \quad n \geq 2.$$

If $f$ is a map $f\colon (I^n, \dot{I}^n) \to (X,x_o)$, $n \geq 1$, set $\phi(f)$ equal to homology class of the cycle $f\colon I^n \to X$. Then $\phi$ induces a homomorphism

$$\phi\colon \pi_n(X,x_o) \to H_n(X) \quad \text{for} \quad n \geq 1.$$

The following diagram commutes

$$\begin{array}{ccccccccc}
\cdots \to & H_n(X,A) & \xrightarrow{\partial} & H_{n-1}(A) & \xrightarrow{i_*} & H_{n-1}(X) & \xrightarrow{j_*} & H_{n-1}(X,A) & \to \cdots \\
& \uparrow \phi & & \uparrow \phi & & \uparrow \phi & & \uparrow \phi & \\
\cdots \to & \pi_n(X,A,x_o) & \xrightarrow{\partial} & \pi_{n-1}(A,x_o) & \xrightarrow{i_*} & \pi_{n-1}(X,x_o) & \xrightarrow{j_*} & \pi_{n-1}(X,A,x_o) & \to \cdots
\end{array}$$

$$\begin{array}{ccccccc}
\cdots \longrightarrow H_2(X,A) & \xrightarrow{\partial} & H_1(A) & \xrightarrow{i_*} & H_1(X) \\
\uparrow \phi & & \uparrow \phi & & \uparrow \phi \\
\pi_2(X,A,x_o & \xrightarrow{\partial} & \pi_1(A,x_o) & \xrightarrow{i_*} & \pi_1(X,x_o)
\end{array}$$

Definition. The pair $(X,A)$ is called n-connected for $n \geq 1$ if

(a) every path component of $X$ intersects $A$

(b) every map $(E^1,S^o) \to (X,A)$ is homotopic relative to $S^o$ to a map of $E^1$ into $A$.

(c) $\pi_k(X,A,x_o) = 0$ for all $2 \leq k \leq n$ and $x_o$ in $A$.

Hurewicz Isomorphism Theorem. Let $A$ be a subspace of $X$ and let $x_o$ be a point in $A$. Assume $A$ and $X$ are arcwise connected and simply connected. If $(X,A)$ is (n-1) connected with $n \geq 2$, then $H_q(X,A) = 0$ for $q < n$ and the map

$$\phi\colon \pi_n(X,A,x_o) \to H_n(X,A) \quad \text{is an isomorphism.}$$

Conversely, if there is an integer $n \geq 2$ such that $H_q(X,A) = 0$ for $q < n$, then $\pi_q(X,A,x_o) = 0$ for $q < n$ and the map

$$\phi\colon \pi_n(X,A,x_o) \to H_n(X,A) \quad \text{is an isomorphism.}$$

Assume $X$ is arcwise connected and simply connected.

If $\pi_q(X,x_o) = 0$ for $q < n$ $(n \geq 2)$, then $H_q(X,x_o) = 0$ for $q < n$ and the map

$$\phi\colon \pi_n(X,x_o) \to H_n(X) \quad \text{is an isomorphism.}$$

If there is an integer $n \geq 2$ such that $H_q(X) = 0$ for $q < n$, then $\pi_q(X,x_o) = 0$ for $q < n$ and the map

$$\phi\colon \pi_n(X,x_o) \to H_n(X) \quad \text{is an isomorphism.}$$

For a more general version of the Hurewicz Isomorphism Theorem see sections 4 and 5 in Chapter 7 of [Spanier].

Theorem 3.2.1. Let $E \xrightarrow{P} B$ be a coordinate bundle with fibre $F$. Let $b_o \in B$. Set $F_o = p^{-1}(b_o)$ and let $x_o \in F_o$. Then $P_*\colon \pi_n(E,F_o,x_o) \to \pi_n(B,b_o)$ is an isomorphism for all $n \geq 2$.

Proof. (a) $P_*\colon \pi_n(E,F_o,x_o) \to \pi_n(B,b_o,b_o)$ is a monomorphism.

Let $\alpha \in \pi_n(E,F_o,x_o)$ with $P_*(\alpha) = 0$. Let the map $f(E^n,S^{n-1},*) \to (E,F_o,x_o)$ be such that $[f] = \alpha$. $p_*(\alpha) = 0$ implies that there is a map $H\colon (E^n \times I,\ S^{n-1} \times I,\ * \times I) \to (B,b_o,b_o)$ so that the diagram

$$\begin{array}{ccc}
(E^n,S^{n-1},*) & \xrightarrow{\quad f \quad} & (E,F_o,x_o) \\
\Big\downarrow {\scriptstyle x \mapsto (x,0)} & & \Big\downarrow {\scriptstyle P} \\
(E^n \times I,\ S^{n-1} \times I,\ * \times I) & \xrightarrow{\quad H \quad} & (B,b_o,b_o)
\end{array}$$

commutes and $H$ restricted to $E^n \times 1$ is the constant map $b_o$. From the second covering homotopy theorem, it follows that there is a stationary lifting

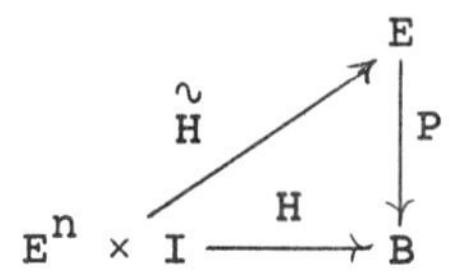


such that $\tilde{H}$ restricted to $E^n \times 0$ is the map f.

> <u>Note</u>: $\tilde{H}$ maps $S^{n-1} \times I$ into $F_o$ by commutivity and $\tilde{H}$ maps $* \times I$ to $x_o$ since $\tilde{H}$ is stationary.

Thus the lifting $\tilde{H}$ yields a commutative diagram

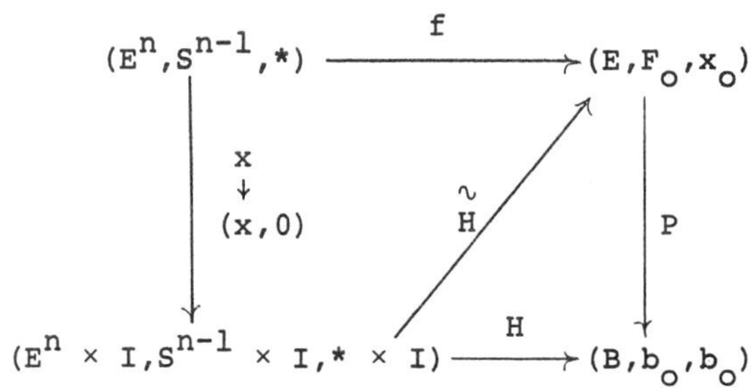


Thus $[f] = [\tilde{H}/E^n \times 1]$. Since H restricted to $E^n \times 1$ is the constant $b_o$, $\tilde{H}$ maps $E^n \times 1$ into $F_o$. So $[\tilde{H}/E^n \times 1] = 0$ in $\pi_n(E,F_o,x_o)$, and this completes the proof that $P_*$ is a monomorphism.

(b) $P_*$ is onto.

Let f be a map $f: (E^n,S^{n-1},*) \to (B,b_o,b_o)$. Define the map

$\sigma\colon S^{n-1} \times I \to E^n$ by $(x,t) \to xt + *(1-t)$. Note $\sigma$ is onto and $E^n$ has the quotient space topology determined the map $\sigma$. Consider

$$\begin{array}{ccccc} & & & \nearrow & x_0 \\ & & & & \downarrow p \\ S^{n-1} \times 0 & \longrightarrow & * & \xrightarrow{f} & b_0 \end{array}$$

From the second covering homotopy theorem, it follows that there is a lifting $\tilde{H}\colon S^{n-1} \times I \to E$ such that

$$\begin{array}{ccccc} & & \tilde{H} & \nearrow & E \\ & & & & \downarrow p \\ S^{n-1} \times I & \xrightarrow{\sigma} & E^n & \xrightarrow{f} & B \end{array}$$

commutes and $\tilde{H}$ is stationary.

Note: $\tilde{H}$ maps $S^{n-1} \times 1$ into $F_0$ by commutivity
$\tilde{H}$ maps $* \times I$ to $x_0$ since $\tilde{H}$ is stationary

In particular $\tilde{H}$ maps $S^{n-1} \times 0 \cup (* \times I)$ to $x_0$.

Since $\sigma$ is a quotient map, there is a map $f'(E^n,S^{n-1},*) \to (E,F_0,x_0)$ such that

$$\begin{array}{ccccc} & & \tilde{H} \;\; f' & \nearrow & (E,F_0,x_0) \\ & & & & \downarrow p \\ (S^{n-1} \times I, S^{n-1} \times 1, * \times I) & \xrightarrow{\sigma} & (E^n,S^{n-1},*) & \xrightarrow{f} & (B,b_0,b_0) \end{array}$$

commutes. Since $p \circ f' = f$, $P_*$ is onto.

Corollary 3.2.2. If $E \xrightarrow{P} B$ is a coordinate bundle with fibre F, then there is a long exact sequence

$$\cdots \longrightarrow \pi_n(F,x_o) \xrightarrow{i_*} \pi_n(E,x_o) \xrightarrow{P_*} \pi_n(B,b_o) \xrightarrow{\partial} \pi_{n-1}(F,x_o) \longrightarrow \cdots$$

$$\cdots \longrightarrow \pi_2(E,x_o) \xrightarrow{P_*} \pi_2(B,b_o) \xrightarrow{\partial} \pi_1(F,x_o) \ .$$

Proof. Replace $\pi_n(E,F,x_o)$ with $\pi_n(B,b_o)$ in the exact homotopy sequence of the pair (E,F).

## 3. Universal Bundles

Given spaces X and B, define the relation $\sim$ on the set of maps $f: X \to B$ by setting $f \sim g$ if there is a map $H: X \times I \to B$ such that $H(x,0) = f(x)$ and $H(x,1) = g(x)$ for all x in X. Then $\sim$ is an equivalence relation. The corresponding set of equivalence classes, denoted by [X,B], is called the set of homotopy classes of maps of X into B.

From Theorem 3.1.3 we know that if $E \xrightarrow{\pi} B$ is a coordinate bundle and f,g are homotopic maps of X into B then the pullbacks $E_f, E_g$ are equivalent coordinate bundles. Thus for a fixed bundle $E \xrightarrow{\pi} B$ with group G, the pullback construction induces a map $[X,B] \to K_G(X)$. In this section we give sufficient conditions on E and X to guarantee that the map above is 1-1 and onto.

Definition. Let $E \xrightarrow{\pi} B$ be a principal G-bundle. The bundle $E$ is called n-universal if for any n-dimensional C. W. complex $K$, subcomplex $L$ of $K$, principal G-bundle $E' \xrightarrow{\pi'} K$, and map

$$\begin{array}{ccc} E'/L & \xrightarrow{\tilde{f}_L} & E \\ \downarrow{\scriptstyle \pi'/L} & & \downarrow{\scriptstyle \pi} \\ L & \xrightarrow{f_L} & B \end{array}$$

of coordinate bundles, there is a map

$$\begin{array}{ccc} E' & \xrightarrow{\tilde{f}} & E \\ {\scriptstyle \pi'}\downarrow & & \downarrow{\scriptstyle \pi} \\ K & \xrightarrow{f} & B \end{array}$$

of coordinate bundles which extends $\tilde{f}_L$.

Theorem 3.3.1. If $E \xrightarrow{\pi} B$ is a principal G-bundle which is (n+1) universal, and $X$ is an n-dimensional C. W. complex, then the map $[X,B] \to K_G(X)$ defined by $[f] \to [E_f]$ is 1-1 and onto.

Proof. (a) $[X,B] \to K_G(X)$ is onto.

Let $E'$ be a principal G-bundle over $X$. Let $x_o$ be a 0-cell in $X$, and let $b$ be a point in $B$. Define $f_{x_o} : x_o \to B$ by $f_{x_o} x_o = b$. Then the pullback $E_{f_{x_o}}$ and the restriction of $E'$ to

$x_o, E'_{x_o}$, are both equivalent to the trivial bundle and hence equivalent. Thus there is a map

$$\begin{array}{ccc} E'_{x_o} & \xrightarrow{\tilde{f}_{x_o}} & E \\ {\scriptstyle \pi'/x_o}\downarrow & & \downarrow{\scriptstyle \pi} \\ x_o & \xrightarrow{f_{x_o}} & B \end{array}$$

of coordinate bundles. From the definition of (n+1) universal bundle, it follows that there is a map of bundles

$$\begin{array}{ccc} E' & \xrightarrow{\tilde{f}} & E \\ \downarrow{\scriptstyle \pi'} & & \downarrow{\scriptstyle \pi} \\ X & \xrightarrow{f} & B \end{array}$$

This shows $[X,B] \to K_G(B)$ is onto.

(b) $[X,B] \to K_G(B)$ is 1-1

Let $f,g$ be maps of $X$ into $B$ such that $E_f$ and $E_g$ are equivalent bundles. Then we must show $f$ and $g$ are homotopic maps of $X$ into $B$.

Since $E_f$ and $E_g$ are equivalent, there is a map

$$\begin{array}{ccc} E_f & \xrightarrow{\tilde{g}} & E \\ \downarrow & & \downarrow \pi \\ X & \xrightarrow{g} & B \end{array}$$

of coordinate bundles. Consider the bundle $E_f \times I$ over $X \times I$. $X \times I$ is an $(n+1)$ dimensional C. W. complex and $(X \times 0) \cup (X \times 1)$ is a subcomplex. Thus the map

$$\begin{array}{ccc} (E_f \times 0) \cup (E_f \times I) & \xrightarrow{\tilde{f} \cup \tilde{g}} & E \\ \downarrow & & \downarrow \pi \\ (X \times 0) \cup (X \times 1) & \xrightarrow{f \cup g} & B \end{array}$$

can be extended to a map of coordinate bundles

$$\begin{array}{ccc} E_f \times I & \xrightarrow{\tilde{H}} & E \\ \downarrow & & \downarrow \pi \\ X \times I & \xrightarrow{H} & B \end{array}$$

$H(x,0) = f(x)$ and $H(x,1) = g(x)$ for all $x$ in $X$. So $f$ and $g$ are homotopic.

<u>Theorem 3.3.2</u>. A principal bundle $E \xrightarrow{\pi} B$ is n-universal iff $E$ is arcwise connected and $\pi_i(E) = 0$ for $1 \leq i < n$.

<u>Proof</u>. Assume $E \xrightarrow{\pi} B$ is n-universal and show that $E$ is arcwise connected and $\pi_i(E) = 0$ for $1 \leq i < n$. (Recall that if $E$ is arcwise connected, $\pi_i(E,e) = \pi_i(E,e')$ for any two points $e$ and $e'$ in $E$.)

Let $f\colon S^i \to E$ be a map, $0 \leq i < n$, and let

$$\begin{array}{ccc} E \times G & \xrightarrow{h} & E \\ \downarrow & & \downarrow{\scriptstyle \pi} \\ E & \xrightarrow{\pi} & B \end{array}$$

be the principal map (see section 5, Chapter 2) and define the map of bundles

$$\begin{array}{ccc} S^i \times G & \xrightarrow{\tilde{f}} & E \\ \downarrow{\scriptstyle P_1} & & \downarrow{\scriptstyle \pi} \\ S^i & \xrightarrow{\pi \circ f} & B \end{array}$$

to be the composition

$$\begin{array}{ccccc} S^i \times G & \xrightarrow{f \times 1_G} & E \times G & \xrightarrow{h} & E \\ \downarrow{\scriptstyle P_1} & & \downarrow{\scriptstyle P_1} & & \downarrow{\scriptstyle \pi} \\ S^i & \xrightarrow{f} & E & \xrightarrow{\pi} & B \end{array} \quad .$$

Note $\tilde{f}(x,e) = f(x)$ for all $x$ in $S^i$. Consider the bundle $E^{i+1} \times G \xrightarrow{P_1} E^{i+1}$. $E^{i+1} \times G$ restricted to $S^i$ is $S^i \times G$. $E$ is n-universal and $i + 1 \leq n$. So $\tilde{f}$ can be extended to a map

$$\begin{array}{ccc} E^{i+1} \times G & \xrightarrow{\tilde{F}} & E \\ \downarrow{\scriptstyle P_1} & & \downarrow{\scriptstyle \pi} \\ E^{i+1} & \xrightarrow{F} & B \end{array}$$

Then $\tilde{F}/E^{i+1} \times e$ is an extension of $f\colon S^i \to E$. Thus for $i$, $0 \leq i < n$ any map $f\colon S^i \to E$ can be extended to a map of $E^{i+1}$ into $E$. For $i = 0$, this shows $E$ is arcwise connected and for $i$, $1 \leq i < n$ we can conclude $\pi_i(E) = 0$.

Assume $E$ is arcwise connected and $\pi_i(E) = 0$ for $1 \leq i < n$. Show that $E$ must be n-universal. First we show that for $0 \leq i < n$; any bundle map

$$\begin{array}{ccc} S^i \times G & \xrightarrow{\tilde{f}} & E \\ \downarrow{\scriptstyle P_1} & & \downarrow{\scriptstyle \pi} \\ S^i & \xrightarrow{f} & B \end{array}$$

can be extended to a bundle mapping

$$\begin{array}{ccc} E^{i+1} \times G & \xrightarrow{\tilde{F}} & E \\ \downarrow & & \downarrow{\scriptstyle \pi} \\ E^{i+1} & \xrightarrow{F} & B \end{array} .$$

Let

$$\begin{array}{ccc} S^i \times G & \xrightarrow{\tilde{f}} & E \\ \downarrow{\scriptstyle p_1} & & \downarrow{\scriptstyle \pi} \\ S^i & \xrightarrow{f} & B \end{array}$$

be a map of principal G-bundles. Since $E$ is arcwise connected and $\pi_i(E) = 0$ for $i$, $1 \leq i < n$. The map $S^i \to E$ defined by $x \to \tilde{f}(x,e)$ can be extended to a map of $E^{i+1} \xrightarrow{P} E$.

Define $E^{i+1} \times G \xrightarrow{\tilde{F}} E$ to be the composition

$$\begin{array}{ccccc} E^{i+1} \times G & \xrightarrow{P \times 1_G} & E \times G & \xrightarrow{h} & E \\ \downarrow{\scriptstyle p_1} & & \downarrow{\scriptstyle p_1} & & \downarrow{\scriptstyle \pi} \\ E^{i+1} & \longrightarrow & E & \xrightarrow{\pi} & B \end{array} .$$

Then

$$\begin{array}{ccc} E^{i+1} \times G & \xrightarrow{\tilde{F}} & E \\ \downarrow{\scriptstyle p_1} & & \downarrow{\scriptstyle \pi} \\ E^{i+1} & \xrightarrow{F} & B \end{array}$$

is a map of coordinate bundles. To see that $\tilde{F}/S^i \times G$ is the same as $f$, pick $x \in S^i$ and consider the equivalence of coordinate bundles.

$$\begin{array}{ccccc} \{x\} \times G & \xrightarrow{\tilde{F}} & \pi^{-1}(f(x)) & \xleftarrow{\tilde{f}} & \{x\} \times G \\ \downarrow{\scriptstyle p_1} & & \downarrow & & \downarrow{\scriptstyle p_1} \\ \{x\} & \xrightarrow{F} & \{f(x)\} & \xleftarrow{f} & \{x\} \end{array}$$

$\{x\} \times G \xrightarrow{\tilde{f}^{-1}\tilde{F}} \{x\} \times G$ is a map of coordinate bundles, so there is an element $g_x$ in $G$ such that $\tilde{f}^{-1}\tilde{F}$ has the form $(x,g) \to (x,g_x g)$ for all $g$ in $G$. In particular, $\tilde{f}^{-1}\tilde{F}(x,e) = (x,g_x)$. But from the definition of $\tilde{F}$, it follows that $\tilde{F}(x,e) = \tilde{f}(x,e)$ for all $x$ in $S^i$. Thus $\tilde{f}^{-1}\tilde{F}(x,g) = (x,g)$ and $\tilde{F}(x,g) = \tilde{f}(x,g)$. So $\tilde{F}$ extends $\tilde{f}$.

Assume $K$ is an n-dimensional C. W. complex, $E' \xrightarrow{\pi'} K$ is a principal G-bundle, $L$ is a sub-complex of $K$ and

$$\begin{array}{ccc} E'/L & \xrightarrow{\tilde{f}} & E \\ {\scriptstyle \pi'/L}\downarrow & & \downarrow{\scriptstyle \pi} \\ L & \xrightarrow{f} & B \end{array}$$

is a map of G-bundles. Since any bundle over a point is equivalent to the trivial bundle, $\tilde{f}$ can be extended to $E'/L \cup K^0$ where $K^0$ is the 0-skeleton of $K$.

Assume $\tilde{f}$ has been extended to $E'/L \cup K^i$ with $0 \leq i < n$. $E'$ restricted to the closure of any $(i+1)$ cell is equivalent to the trivial bundle, so the restriction of $E'$ to the boundary of any $(i+1)$ cell must also be equivalent to the trivial bundle. Using the remarks above, we can extend $\tilde{f}$ restricted to the boundary of any $(i+1)$ cell of $K$ not in $L$, to the restriction of $E'$ to the $(i+1)$ cell. In this way $\tilde{f}$ can be extended to $E'/L \cup K^{i+1}$. This completes the proof.

<u>Definition</u>. If $G$ is a topological group with identity $e$, set $\pi_o(G,e)$ equal to the set of homotopy classes of maps of pairs, $[(S^o,1),(G,e)]$. Given maps $f$ and $g\colon (S^o,1) \to (G,e)$ define $f\cdot g(x)$ to be $f(x)g(x)$, then $f \times g \to fg$ gives $\pi_o(G,e)$ the structure of a group.

<u>Lemma 3.3.3</u>. If $E \xrightarrow{\pi} B$ is an n-universal principal G-bundle, with $n \geq 2$, then $\pi_1(B,b_o)$ is isomorphic to $\pi_o(G,e)$.

<u>Proof</u>. Define $\partial\colon \pi_1(B,b_o) \to \pi_o(G,e)$ as follows: Let $f$ be a map $f\colon (I,S^o) \to (B,b_o)$. From the second covering homotopy theorem, it follows that there is a map $\tilde{f}_1\colon I \to E$ such that

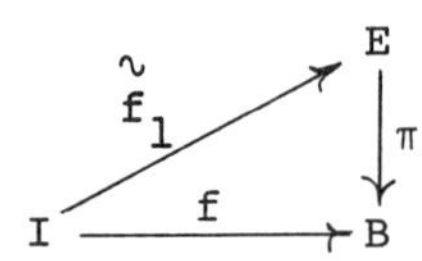


commutes and $\tilde{f}_1(1) = e$. From the commutivity of the diagram, it follows that $\tilde{f}_1$ maps $S^o$ into $G$. Set $\partial f$ equal to the map $\tilde{f}_1$ restricted to $S^o$. To see that the element $[\tilde{f}_1]$ in $\pi_o(G,e)$ is independent of the lifting $\tilde{f}_1$, suppose $\tilde{f}_2$ is another lifting and define $F\colon I \times I \to B$ by

$$F(t_1,t_2) = \begin{cases} f(2t_1) & 0 \leq t_1 \leq \frac{1}{2}(1-t_2) \\ f(t_2) & \frac{1}{2}(1-t_2) \leq t_1 \leq 1 - \frac{1}{2}(1-t_2) \\ f(2-2t_1) & 1 - \frac{1}{2}(1-t_2) \leq t_1 \leq 1 \end{cases}$$

and $\tilde{F}: I \times 0 \to E$ by

$$\tilde{F}(t,0) = \left\{ \begin{array}{ll} \tilde{f}_1(2t) & 0 \le t \le \frac{1}{2} \\ \tilde{f}_2(2-2t) & \frac{1}{2} \le t \le 1 \end{array} \right\}.$$

From the second covering homotopy theorem, it follows that $\tilde{F}$ can be extended to $I \times I$ such that

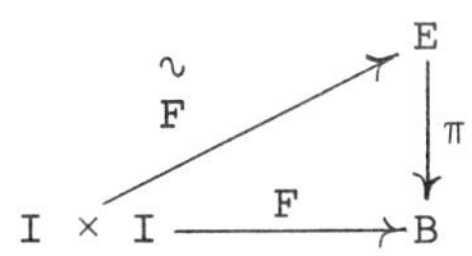


commutes and $\tilde{F}$ is stationary. Since $\tilde{F}$ is stationary $\tilde{F}(1,1) = \tilde{F}(1,0) = \tilde{f}_2(0)$ and $\tilde{F}(0,1) = \tilde{F}(0,0) = \tilde{f}_1(0)$ so $\tilde{F}$ restricted to $I \times 1$ is a path in $G$ connecting $\tilde{f}_2(0)$ and $\tilde{f}_1(0)$. It follows that $[\tilde{f}_1/S^o]$ equals $[\tilde{f}_2/S^o]$ in $\pi_o(G,e)$.

From the second covering homotopy theorem and the fact that $\partial f$ is independent of the choice of lifting $\tilde{f}$, it follows that $\partial$ induces a homomorphism of $\pi_1(B,b_o)$ into $\pi_o(G,e)$. That $\partial$ is a monomorphism follows from the assumption that $\pi_1(E,e) = 0$. $\partial$ is onto because $E$ is arcwise connected.

<u>Examples</u>. Since the Stiefel Manifold $G(n,k)$ is arcwise connected and $\pi_i(G(n,k),I) = 0$ for $i < (k+1)d-1$ (see previous section), the

principal $G(n)$ bundle $G(n+k,k) \xrightarrow{\pi} M(n+k,k)$ is $(k+1)d-1$ universal. Moreover, if $H$ is any closed subgroup of $G(n)$, then $H \times G(k)$ admits a local section in $G(n+k)$ for all values of $k$. [See Chevally-Theory of Lie Groups, Proposition 1, page 110]. From the bundle structure theorem for Homogeneous spaces, it follows that for any closed subgroup $H$ of $G(n)$

$$\begin{array}{c} G(n+k,k) \\ \downarrow \pi \\ G(n+k)/H \times G(k) \end{array}$$

is a $(k+1)d-1$ universal principal H-bundle. Consider the special case $d = 1$, $n = 1$, then we have that

$$\begin{array}{c} G(k+1,k) = S^k \\ \downarrow \\ M(k+1,k) = RP(k), \end{array}$$

is a k-universal principal $Z_2$ bundle. From Whitehead's cellular approximation theorem it follows that if $X$ is a finite dimensional C. W. complex, then the usual embedding of $RP(N)$ in $RP(N+1)$ induces a 1-1 onto map $[X,RP(N)] \to [X,RP(N+1)]$ for large values of $N$. Thus $K_{Z_2}(X)$ can be identified with $[X,RP(\infty)]$ for any finite C.W. complex $X$.

From the lemma above, and the exact homotopy sequence for coordinate bundles, it follows that

$$\pi_i(RP(\infty)) = \begin{cases} Z_2 & i = 1 \\ 0 & \text{otherwise} \end{cases}.$$

Thus $RP(\infty)$ is an Eilenberg-MacLane space $K(Z_2,1)$. Hence there is a natural equivalence of functors $[X,RP(\infty)] \to H^1(X\colon Z_2)$ for C. W. complexes X. (See [Spanier]) Thus for finite dimensional C. W. complexes $X, K_{Z_2}(X)$ can be identified with the first cohomology group of X with $Z_2$ coefficients.

Similarly, for any finite dimensional C. W. complex $X, K_{S^1}(X)$ can be identified with $[X,CP(\infty)]$ and

$$\pi_i(CP(\infty)) = \begin{cases} Z & \text{for } i = 2 \\ 0 & \text{otherwise} \end{cases}.$$

$CP(\infty)$ is an Eilenberg-MacLane Space $K(Z,2)$ and hence $K_{S^1}(X)$ can be identified with $H^2(X\colon Z)$ for any finite dimensional C. W. complex X.

Let $S^3$ denote the group of Quaternions of norm 1. For any finite dimensional C. W. complex $X, K_{S^3}(X)$ can be identified with $[X,QP(\infty)]$ where $QP(\infty)$ denotes infinite dimensional Quaternionic projective space. Note $\pi_i(QP(\infty)) = \pi_{i-1}(S^3,e)$ for all $i \geq 1$.

## 4. Milgram's Construction

The purpose of this section is to sketch Milgram's construction of the universal classifying space, $B_G$, of a topological group, G. Two important properties of Milgram's classifying space are that $B_{G\times H}$ is homeomorphic to $B_G \times B_H$ and that $B_G$ is an abelian topological group if G is abelian. The reader is referred to [Steenrod 3] for details.

In order to avoid various difficulties, it is necessary to restrict to the category of compactly generated spaces. The Hausdorff space X is called compactly generated if each subset that intersects every compact set in a closed set is itself a closed set. Given a Hausdorff space X, k(X) denotes the set X with topology defined by: $A \subseteq k(X)$ is closed iff $A \cap C$ is closed in X for every compact subset C of X. The product of two compactly generated spaces X and Y is by definition $k(X \times Y)$ where $X \times Y$ is the usual product. $k(X \times Y)$ will be denoted simply by $X \times Y$. In the category of compactly generated spaces, a pair (X,A) is a compactly generated space X and a closed subspace A such that (X,A) is an NDR. That is there are maps $u: X \to I$ and $h: X \times I \to X$ such that

1. $A = u^{-1}(0)$
2. $h(x,0) = x$ for x in X
3. $h(x,t) = x$ for $(x,t)$ in $A \times I$

and

4. $h(x,1) \in A$ for x in $u^{-1}[0,1]$.

Two of the advantages of restricting to the category of compactly generated spaces are that the product of quotient maps is a quotient map and that if one forms the union of an expanding sequence of spaces, in which each successive pair is an NDR, and gives the union the weak topology then the usual difficulties (e.g. Hausdorff, behavior under products, etc.) disappear. See [Steenrod 4] for details. Without further comment, we will assume that all topological groups are compactly generated and that if $G$ is a topological group, then $(G,e)$ in an NDR.

<u>Notation</u>: If $A$ is a subspace of $X$ and $f$ is a map of $A$ into $B$, then the quotient space of the disjoint union, $X \cup B$, obtained by identifying $x$ with $f(x)$ for all $x$ in $A$ will be denoted by $X \cup_f B$.

Let $G$ be a topological group. A space $A$ is called a right G-space if there is a map $h: A \times G \to A$ such that $h(x,e) = x$ for all $x$ in $A$ and $h(h(x,g_1),g_2) = h(x,g_1g_2)$. $h$ is called an action of $G$ on $A$.

<u>Definition</u>. Let $X$ be a space, $A$ a closed subspace, $\phi: A \times G \to A$ an action of $G$ on $A$. Define the enlargement to $X$ of the G-action on $A$ to be the space $\overline{X} = X \cup_\phi A$.

The map $\overline{X} \times G \to \overline{X}$ defined by $([x,g_1],g_2) \to [x,g_1g_2]$ extends the action of $G$ on $A$ to an action of $G$ on $\overline{X}$.

Example: Recall

$$G(1) = \begin{cases} Z_2 & d = 1 \\ S^1 & d = 2 \\ S^3 & d = 4 \end{cases}$$

$$E^{nd} = \{(Z_1, \ldots Z_n) \text{ in } F^n\colon \Sigma\, \overline{Z}_i Z_i \leq 1\}$$

$$S^{nd-1} = \{(Z_1, \ldots Z_n) \text{ in } F^n\colon \Sigma\, \overline{Z}_i Z_i = 1\}.$$

The map $S^{nd-1} \times G(1) \to S^{nd-1}$ defined by $(Z_1, \ldots Z_n), g)$ $\to (Z_1 g, \ldots, Z_n g)$ is a $G(1)$ action on $S^{nd-1}$. Set $\overline{E}^{nd}$ equal to the enlargement of the $G(1)$ action on $S^{nd-1}$ to $E^{nd}$. View $\overline{E}^{nd}$ as the quotient space of $E^{nd} \times G(1)$ obtained by identifying $(Z_1, \ldots Z_n, g)$ with $(Z_1 g, \ldots Z_n g, e)$ if $\Sigma\, \overline{Z}_i Z_i = 1$, and define $f\colon (\overline{E}^{nd}, S^{nd-1}) \to (S^{(n+1)d-1}, S^{nd-1})$ by $f[Z_1, \ldots Z_n, g]$ $= (z_1 g, \ldots Z_n g, (\sqrt{1 - \Sigma\, \overline{Z}_i Z_i})g)$. Then $f$ is a homeomorphism of $\overline{E}^{nd}$ onto $S^{(n+1)d-1}$, and the $G(1)$ action on $\overline{E}^{nd}$ corresponds to the usual $G(1)$ action on $S^{(n+1)d-1}$.

Lemma 3.4.1: (a) $\overline{X}$ is the quotient space of $X \times G$ by the relations $(x,g) = (\phi(x,g), e)$ for all $x$ in $A$, $g$ in $G$.

(b) $\overline{X} - A$ is homeomorphic to $(X-A) \times G$, and

(c) $\overline{X} - X$ is homeomorphic to $(X-A) \times (G-e)$.

Definition. Let $X$ be a space, $x_o$ a point in $X$. Set $X \wedge I$ equal to $(X \times I) \cup_f (x_o)$ where $f$ is the map $f\colon (X \times 0) \cup (x_o \times I) \to x_o$.

Roughly speaking, $X \wedge I$ is obtained from $X$ by adjoining a path from each point $x$ in $X$, $x \neq x_o$, to the point $x_o$.

Definition. A contraction of the pointed space $(X,x_o)$ to the point $x_o$ is a map $h$ of $X \wedge I$ into $X$ such that

$$h[x,1] = x \quad \text{for all } x \text{ in } X$$

$$h[h[x,t_1],t_2] = h[x,t_1t_2]$$

and

$$h[x_o,0] = x_o.$$

The contraction of $X \wedge I$ to $x_o$ defined by $([x,t_1],t_2) \to [x,t_1t_2]$ is called the canonical contraction of $X \wedge I$.

Definition. Let $A$ be a closed subspace of $X$ and let $x_o$ be a point in $A$. Assume $h\colon A \wedge I \to A$ is a contraction. Define the enlargement to $X$ of the contraction on $A$ to be the space $\tilde{X}$ defined by $\tilde{X} = (X \wedge I) \cup_h A$. The contraction $\tilde{h}\colon \tilde{X} \wedge I \to \tilde{X}$ defined by $([x,t_1],t_2) \to [x,t_1t_2]$ restricts to $h$ on $A \wedge I$.

Roughly speaking, $\tilde{X}$ is obtained from $X$ by adjoining a path from $x$ to $x_o$, for each point $x$ in $(X-A)$.

Example. If $X = S^n$, $A = \{(x_1 \cdots x_{n+1})$ in $S^n$: $x_{n+1} \geq 0\}$, $x_o = (1,0\cdots 0)$, $h: A \wedge I \to A$ a contraction; then $\tilde{X}$ can be identified with $E^{n+1}$.

Lemma 3.4.2. (a) $X$ is the quotient of $X \times I$ by the relations

(1) $(x,0) = (x_o,1)$ for all $x$ in $X$

$[(x_o,t) = (x_o,1)$ for all $t$ in $I]$

(2) $(x,t) = (h(x,t),1)$ for all $x$ in $A$, $t$ in $I$. Note: the relation [ ] is implied by 2.

(b) $\tilde{X} - A$ is homeomorphic to $(X-A) \times (I-0)$

(c) $\tilde{X} - X$ is homeomorphic to $(X-A) \times (I-\dot{I})$ where $\dot{I} = \{0,1\}$.

Define the spaces $D_i$ and $E_i$ for $i \geq 0$ by induction as follows. Set

$D_o = \{pt\}$

$E_o = G$

$D_1 = G \wedge I$

$E_1$ = enlargement to $D_1$ of the $G$ action on $E_o$

$D_2$ = enlargement to $E_1$ of the contraction on $D_1$.

In general

$$E_n = \text{enlargement to } D_n \text{ of the G-action on } E_{n-1}$$

$$D_{n+1} = \text{enlargement to } E_n \text{ of the contraction on } D_n.$$

There are inclusions

$$D_o \hookrightarrow E_o \hookrightarrow D_1 \hookrightarrow E_1 \hookrightarrow \cdots\cdots \hookrightarrow E_{n-1} \hookrightarrow D_n \hookrightarrow E_n \hookrightarrow D_{n+1} \hookrightarrow \cdots\cdots$$

where $D_o \hookrightarrow E_o$ is given by $pt \hookrightarrow e$ and $E_o \hookrightarrow D_1$ is given by $g \hookrightarrow [g,1]$.

By construction each $D_n$ has a contraction $h_n \colon D_n \wedge I \to D_n$ such that $h_n$ restricted to $D_{n-1}$ is $h_{n-1}$, and each $E_n$ has a G-action $\phi_n \colon E_n \times G \to E_n$ such that $\phi_n$ restricted to $E_{n-1}$ is $\phi_{n-1}$.

Set $E_G = \bigcup_{i \geq 0} E_i = \bigcup_{i \geq 0} D_i$ with the weak topology. $E_G$ is contractible and a G-space. Set $B_G$ equal to the orbit space $(E_G)/G$ with the quotient space topology.

<u>Examples</u>.

1. $G = Z_2$.

$$\begin{array}{ccccccccccccccc} D_o & \hookrightarrow & E_o & \hookrightarrow & D_1 & \hookrightarrow & E_1 & \hookrightarrow \cdots \hookrightarrow & E_n & \hookrightarrow & D_{n+1} & \hookrightarrow & E_{n+1} & \hookrightarrow & \cdots \\ \| & & \| & & \| & & \| & & \| & & \| & & \| & & \\ pt & \hookrightarrow & S^o & \hookrightarrow & I & \hookrightarrow & S' & \hookrightarrow \cdots \hookrightarrow & S^n & \hookrightarrow & E^{n+1} & \hookrightarrow & S^{n+1} & \hookrightarrow & \cdots \end{array}$$

The $Z_2$ action on $E_n$ corresponds to the antipodal map on $S^n$. $B_{Z_2} \cong RP(\infty)$.

2. $G = S^1$

$$\begin{array}{ccccccccccccccc} D_0 & \hookrightarrow & E_0 & \hookrightarrow & D_1 & \hookrightarrow & E_1 & \hookrightarrow & \cdots\cdots \hookrightarrow & E_{n-1} & \hookrightarrow & D_n & \hookrightarrow & E_n & \hookrightarrow \cdots \\ \| & & \| & & \| & & \| & & & \| & & \| & & \| & \\ pt & \hookrightarrow & S^1 & \hookrightarrow & E^2 & \hookrightarrow & S^3 & \hookrightarrow & \cdots \hookrightarrow & S^{2n-1} & \hookrightarrow & E^{2n} & \hookrightarrow & S^{2n+1} & \hookrightarrow \cdots \end{array}$$

The $S^1$ action on $E_n$ corresponds to the usual action of $S^1$ on $S^{2n+1}$. $B_{S^1} \cong \mathbb{C}P(\infty)$.

3. $G = S^3$

$$\begin{array}{ccccccccccccccc} D_0 & \hookrightarrow & E_0 & \hookrightarrow & D_1 & \hookrightarrow & E_1 & \hookrightarrow & \cdots\cdots \hookrightarrow & E_{n-1} & \hookrightarrow & D_n & \hookrightarrow & E_n & \hookrightarrow \cdots \\ \| & & \| & & \| & & \| & & & \| & & \| & & \| & \\ pt & \hookrightarrow & S^3 & \hookrightarrow & E^4 & \hookrightarrow & S^7 & \hookrightarrow & \cdots\cdots \hookrightarrow & S^{4n-1} & \hookrightarrow & E^{4n} & \hookrightarrow & S^{4n+3} & \hookrightarrow \cdots \end{array}$$

The action of $S^3$ on $E_n$ corresponds to the usual action of $S^3$ on $S^{4n+3}$. $B_{S^3}$ is homeomorphic to $QP(\infty)$, infinite dimensional Quaternionic Projective Space.

From Lemma 3.4.2 with $A = D_{n-1}$, $X = E_{n-1}$, it follows that $D_n - E_{n-1}$ is homeomorphic to $(E_{n-1}-D_{n-1}) \times (I-\dot{I})$.

From Lemma 3.4.1 with $A = E_{n-1}$ and $X = D_n$, it follows that $E_n - D_n$ is homeomorphic to $(D_n-E_{n-1}) \times (G-e)$.

Thus

$$D_o = pt$$

$$E_o - D_o = (G-e)$$

$$(D_n - E_{n-1}) \text{ is homeomorphic to } (G-e)^n \times (I-\dot{I})^n$$

and

$$(E_n - D_n) \text{ is homeomorphic to } (G-e)^{n+1} \times (I-\dot{I})^n.$$

From part (a) of Lemma 3.4.1 and part (a) of Lemma 3.4.2, it follows that $E_n$(resp $D_n$) can be viewed as the quotient space of $G^{n+1} \times I^n$(resp $G^n \times I^n$) by the relations

C1. $(e,t,\ldots) = (e,1,\ldots)$

$(g_1,t_1,\ldots g_{i-1},t_{i-1},e,t_i,g_{i+1}\cdots) = (g_1,t_1,\ldots g_{i-1},t_{i-1}t_i,e,1,g_{i+1},\ldots)$

C2. $(g_1,t_1\cdots g_i,0,g_{i+1},t_{i+1}\cdots) = (e,1,e,1\cdots e,1,g_{i+1},t_{i+1},\cdots)$.

C3. $(g_1,t_1\cdots g_{i-1},1,g_i,t_i\cdots) = (g_1,t_1\cdots,g_{i-1}g_i,1,e,t_i\cdots)$.

A tuple $(g_1,t_1\cdots)$ is said to be in normal form if it has the form

$$(g_1,t_1\cdots,g_i,t_i,e,1,e,1\cdots)$$

with $(g_1,t_1\cdots,g_i,t_i)$ in $(G-e)^i \times (I-\dot{I})^i$

or

$$(g_1, t_1 \ldots, t_{i-1}, g_i, 1, e, 1, e \cdots) \quad \text{with} \quad (g_1, t_1 \ldots t_{i-1}, g_i) \quad \text{in} \quad (G-e)^i \times (I-\dot{I})^{i-1} .$$

It follows that each element in $E_n$ or $D_n$ has a unique representation in normal form.

$D_n$ can also be viewed as a quotient space of $G^n \times \Delta_n$ as follows.

Set

$$\Delta_n = \{(t_1 \cdots t_n) : 0 \leq t_1 \leq t_2 \leq \cdots \leq t_n \leq 1\}$$

$$\delta_n = \{(t_1 \cdots t_n) : 0 < t_1 < t_2 < \ldots < t_n < 1\} = \text{interior of } \Delta_n .$$

Define $\psi_n : I^n \to \Delta_n$ by

$$(t_1 \cdots t_n) \to (t_1 t_2 \cdots t_n, t_2 t_3 \cdots t_n, \ldots, t_i t_{i+1} \cdots t_n \ldots, t_{n-1} t_n, t_n) .$$

Note $\psi_n$ maps the interior of $I^n$ homeomorphically onto $\delta_n$. For $n \geq 1$ define relations on $G^n \times \Delta_n$ by

S1. $[g_1, t_1 \cdots, g_i, t_i, e, t_{i+1}, g_{i+2}, \cdots g_n, t_n]$

$= [g_1, t_1 \cdots g_i, t_i, g_{i+2}, t_{i+2} \cdots g_n, t_n, e, 1]$

$[e, t] = [e, 1]$.

S2. $[g_1, 0, g_2, t_2, \ldots g_n, t_n] = [g_2, t_2 \cdots g_n, t_n, e, 1]$ .

S3. $[g_1, t_1 \cdots, g_i t_i, g_{i+1}, t_i, g_{i+2}, \ldots, g_n, t_n]$

$$= [g_1, t_1 \cdots, g_i g_{i+1}, t_i, g_{i+2}, t_{i+2}, \ldots, g_n, t_n, e, 1] .$$

The tuple $[g_1, t_1 \cdots, g_n, t_n]$ is said to be in normal form if it has the form

$$[g_1, t_1, \ldots, g_i, t_i, e, 1, e, 1 \cdots e, 1]$$
$$\text{with} \quad (g_1, t_1, \ldots g_i, t_i) \; \varepsilon \; (G-e)^i \times \delta_i$$

or

$$[g_1, t_1, \ldots t_{i-1}, g_i, 1, e, 1, e \cdots 1, e, 1]$$
$$\text{with} \quad (g_1, t_1 \cdots t_{i-1}, g_i) \quad \text{in} \quad (G-e)^i \times \delta_{i-1} .$$

Each element of $D_n$ has a unique representation in normal form. For a detailed proof of Theorem 3.4.3 see [Steenrod 3].

Theorem 3.4.3. (a) The quotient space of $G^n \times \Delta_n$ by the relations S1 - S3 is homeomorphic to $D_n$. If $k_n\colon G^n \times \Delta_n \to D_n$ is the quotient map, then

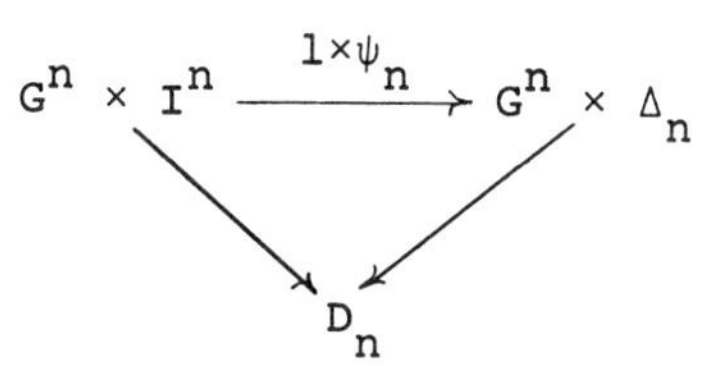

commutes and the restriction of $k_n$ to $G^{n-1} \times \Delta_{n-1}$ is $k_{n-1}$.

(b) Each $k_n$ restricts to homeomorphisms

$$(G-e)^n \times \delta_n \to D_n - E_{n-1}$$

and

$$(G-e)^n \times \delta_{n-1} \to E_{n-1} - D_{n-2} .$$

(c) The restriction of $k_n$ to $G^n \times \Delta_{n-1} \to E_{n-1}$ is a G-map where $g$ acts only on the $n^{th}$ G factor by right translation.

(d) The contraction $h_n\colon D_n \wedge I \to D_n$ is given by

$$([g_1,t_1,g_2,t_2,\ldots g_n,t_n],t) \to [g_1,t_1t,g_2,t_2t,\ldots g_n,t_nt] .$$

(e) Given topological groups $G$ and $H$, then the map $(G \times H)^n \times \Delta_n \to (G^n \times \Delta_n) \times (H^n \times \Delta_n)$ defined by

$$((g_1,h_1),t_1,(g_2,h_2),t_2\ldots,(g_n,h_n),t_n) \to (g_1,t_1g_2,t_2\ldots g_n,t_n) \times (h_1,t_1,h_2,t_2,\ldots h_n,t_n)$$

induces a homeomorphism, $\xi$, of $E_{G\times H}$ onto $E_G \times E_H$.

Next we sketch the proof that $E_G$ is a topological group and

$$\begin{array}{c} E_G \\ \big\downarrow \\ B_G \end{array}$$ is a principal G-bundle.

Given G, let $\tilde{\hat{E}}$ be the free associative monoid generated by pairs $(g,t)$ in $G \times I$, with unit $(\ )$. Define relations on $\tilde{E}$ by

1. $(g,0) = (e,t) = (\ )$
2. $(g,t)(g',t) = (gg',t)$
3. If $0 < t' < t \leq 1$ then $(g,t)(g',t') = (gg'g^{-1},t')(g,t)$.

Let $E_G'$ denote the corresponding set of equivalence classes. Then $E_G'$ is a group. The inverse of $(g,t)$ is $(g^{-1},t)$. If G is abelian, then $E_G'$ is abelian. A monomial $(g_1,t_1) \cdots (g_n,t_n)$ is siad to be in normal form if $0 < t_1 < t_2 \cdots < t_n \leq 1$, and each $g_i$ is in $G - e$. From the relations 1 - 3 above, it follows that each monomial is equivalent to at least one monomial in normal form. To see that each element in $E_G'$ is uniquely represented by a monomial in normal form; associate with each monomial $m = (g_1,t_1) \cdots (g_n,t_n)$ the function $f_m$: $[0,1] \to G$ defined by $f_m(t) = \prod_{t_i > t} g_i$. For example, if $m = (g_1,t_1)(g_2,t_2)(g_3,t_3)$ with $0 < t_1 < t_3 < t_2 < 1$ then

$$f_m(t) = \left\{ \begin{array}{ll} g_1 g_2 g_3 & 0 \leq t < t_1 \\ g_2 \circ g_3 & t_1 \leq t < t_3 \\ g_2 & t_3 \leq t < t_2 \\ e & t_2 \leq t \leq 1 \end{array} \right\} .$$

Given the function $f_m$: $[0,1] \to G$, a monomial in normal form, $m'(f_m)$, can be constructed such that if $m$ is in normal form, then $m'(f_m) = m$ in $\tilde{E}$. It follows that the elements of $E'_G$ are in a 1-1 correspondence with monomials in normal form.

Theorem 3.4.4. (a) Define $k: E'_G \to E_G$ by $(g_1,t_1) \cdots (g_n,t_n) = m \to k_n[g_1,t_1 \cdots g_n,t_n]$ if $m$ is in normal form. Then $k$ is 1-1 and onto.

(b) $E_G$, with multiplication induced by the multiplication on $E'_G$ under $k$, is a topological group.

(c) Under the inclusion $G \to E_G$ defined by $g \to [g,1]$, $G$ becomes a closed subgroup of $E_G$.

(d) If $G \to H$ is a continuous homomorphism, then $E_G \to E_H$ is a continuous homomorphism.

(e) $\xi: E_{G\times H} \to E_G \times E_H$ is an isomorphism of topological groups, and $\xi$ induces a homeomorphism $\xi': B_{G\times H} \to B_G \times B_H$.

(f) If $G$ is abelian, then $E_G$ (and hence $B_G$) are abelian topological groups.

If $X$ is a G-space and $A$ is closed under the G-action, then $(X,A)$ is called a G-NDR if $(X,A)$ can be represented as an NDR by maps $u$ and $h$ with the additional properties that $u(xg) = u(x)$ for all $x$ and $g$ and $h(xg,t) = [h(x,t)]g$ for all $x,g$, and $t$.

It can be shown that since $(G,e)$ is an NDR the pairs $(E_n,G)$ and $(E_G,E_n)$ are G-NDR's for all $n \geq 0$. In particular $(E_G,G)$ is a G-NDR.

<u>Theorem 3.4.5</u>. $E_G \xrightarrow{P} B_G$ is a principal G-bundle.

<u>Proof</u>. Let $u\colon E_G \to I$ and $h\colon E_G \times I \to E_G$ represent $(E_G, G)$ as a G-NDR. Set $W = u^{-1}[0,1)$ and define $r\colon W \to G$ by $r(y) = h(y,1)$. Then $W$ is open in $E_G$ and closed under the action of $G$ on $E_G$. Thus $V = p(W)$ is open in $B_G$. Note $r(yg) = r(y)g$. Define $\xi\colon W \times G \to W$ by $(y,g) \to y(ry)^{-1}g$.

<u>Claim</u>: There is a continuous map $\rho\colon V \times G \to W$ such that

$$\begin{array}{ccc} W \times G & & \\ \downarrow{\scriptstyle P\times 1} & \searrow{\scriptstyle \xi} & \\ V \times G & \xrightarrow{\rho} & W \end{array}$$

commutes. The map $\rho$ exists since

$$\xi(yg,g') = (yg)r(yg)^{-1}g' = yg[r(y)g]^{-1}g' = yr(y)^{-1}g' = \xi(y,g').$$

$\rho$ is continuous since $V \times G$ has the quotient space topology induced by the map $P \times 1$.

Consider the diagram

$$\begin{array}{ccccc} W \times G & & & & \\ {\scriptstyle P\times 1}\big| & \searrow{\scriptstyle \xi} & & & \\ V \times G & \xrightarrow{\rho} & W & \xrightarrow{(p,r)} & V \times G \end{array}$$

It is readily seen that $(p,r)\circ\rho = 1_{V\times G}$ and $\rho\circ(p,r) = 1_W$, and the result follows from the Bundle Structure Theorem for Homogeneous Spaces.

Corollary 3.4.6. Let $f\colon H \to G$ be a continuous homomorphism of topological groups. Assume $(H,e)$ and $(G,e)$ are NDR's and $f_*\colon \pi_n(H,e) \to \pi_n(G,e)$ is an isomorphism for all $n \geq 0$. Then for any C. W. complex $X$, the map $[X,B_H] \xrightarrow{(Bf)} [X,B_G]$ is 1-1 and onto.

Proof. It follows from the commutative diagram

$$\begin{array}{ccc} H & \xrightarrow{f} & G \\ \downarrow & & \downarrow \\ E_H & \xrightarrow{Ef} & E_G \\ \downarrow & & \downarrow \\ B_H & \xrightarrow{Bf} & B_G \end{array}$$

and the corresponding map of the exact homotopy sequences that $(Bf)_*\colon \pi_n(B_H,*) \to \pi_n(B_G,*)$ is an isomorphism for all $n$. The proof is completed by using the fact that for any space $Y$ there is a C. W. complex, $\hat{Y}$, and a map $g\colon \hat{Y} \to Y$ inducing an isomorphism of all homotopy groups. It follows that for any C. W. complex $X$, $[X,\hat{Y}] \xrightarrow{g} [X,Y]$ is 1-1 and onto.

Corollary 3.4.6 can be applied to the inclusion of $O(n)$ in $GL(n,R)$ to show that for any finite C. W. complex $X$, the map

$K_{O(n)}(X) \to K_{GL(n,R)}(X)$ induced by the inclusion of $O(n)$ in $GL(n,R)$ is 1-1 and onto.

<u>Corollary 3.4.7</u>. Assume $(G,e)$ is an NDR, $H$ is a closed subgroup of $G$, and $H$ admits a local section in $G$; then

(a) $$\begin{array}{c} E_G \\ \downarrow \\ (EG)/H = \tilde{B}_H \end{array}$$ is a principal H-bundle, and hence is universal.

(b) $$\begin{array}{c} \tilde{B}_H \\ \downarrow \\ B_G \end{array}$$ is a bundle with fibre $H$, and group $G/H_o$.

(c) Let $X$ be a finite C. W. complex and let $E \xrightarrow{\pi} X$ be a principal $G$ bundle classified by the map $f: X \to B_G$. Then the structure group of $E$ can be reduced to $H$ iff there is a lifting

$$\begin{array}{ccc} & \overset{\tilde{f}}{\nearrow} & \tilde{B}_H \\ & & \downarrow \\ X & \xrightarrow{\ f\ } & B_G \end{array}$$

which, in turn, is equivalent to the existence of a cross section to the pullback, $E_f$, defined by

$$\begin{array}{ccc} E_f & \longrightarrow & \tilde{B}_H \\ \downarrow & & \downarrow \\ X & \xrightarrow{f} & B_G \end{array}$$

(See exercise 7 at the end of Chapter 2).

## 5. $C_*(B_G)$ and the Bar Construction

The purpose of this section is to show that if $G$ is a C. W. complex with cellular multiplication, then $B_G$ is a C. W. complex and $C_*(B_G)$ is isomorphic as a chain complex, to the reduced bar construction applied to $C_*(G)$.

Lemma 3.5.1. If $X$ and $B$ are C. W. complexes, $A$ is a subcomplex of $X$ and the map $f: A \to B$ is cellular, then $X \cup_f B$ can be given the structure of a C. W. complex such that the quotient map $q: X \cup B \to X \cup_f B$ is cellular; $B$ is a subcomplex of $X \cup_f B$; and for each integer $n \geq 0$, the n-cells of $X \cup_f B$ which are not cells of $B$ are in a 1-1 correspondence with the n-cells of $X$ which are not cells of $A$.

Proof. Define the C. W. complexes $Y_i$ inductively as follows. Set $Y_o$ equal to the disjoint union of the C. W. complex $B$ and the discrete space $(X^o - A^o)$.

Define $f^o: X^o \cup A \to Y_o$ by

$$f^o(x) = \begin{Bmatrix} x & \text{if} & x \in X^o - A^o \\ f(x) & \text{if} & x \in A \end{Bmatrix}.$$

Assume the C. W. complex $Y_i$, and cellular map $f^i: X^i \cup A \to Y_i$ have been defined. Define $Y_{i+1}$ and $f^{i+1}$ as follows. Let $\{e_\lambda^{i+1}\}_{\lambda\varepsilon\Lambda}$ denote the i+1-cells of $X^{i+1}$ which are not cells of the subcomplex A, and let $\{f_\lambda\}_{\lambda\varepsilon\Lambda}$ denote the corresponding attaching maps $f_\lambda: S_\lambda^i \to X^i$. Adjoin cells $\{a_\lambda^{i+1}\}_{\lambda\varepsilon\Lambda}$ to $Y_i$ by the attaching maps $f^i f_\lambda$. Then $f^i$ extends to a cellular map $f^{i+1}: X^{i+1} \cup A \to Y_{i+1}$. This defines the sequence of C. W. complexes:

$$B \subseteq Y_o \subseteq Y_1 \subseteq \cdots \subseteq Y_i \subseteq \cdots \text{ and cellular maps } f^i: X^i \cup A \to Y_i.$$

Given $Y = \bigcup_{i \geq 0} Y_i$ the weak topology. Then $\{f^i\}_{i \geq 0}$ induces a cellular map $q: X \cup B \to Y$ which is onto. That $Y$ has the quotient space topology is proved by showing that if $g^{-1}(K)$ is a closed subset of $X \cup B$, then $K$ must be closed in $Y$.

<u>Conventions</u>. Unless otherwise specified, R will denote any commutative ring with unit. If X is a C. W. complex, $C_n(X: R)$ denotes $H_n(X^n, X^{n-1}: R)$ and $C_*(X: R)$ denotes the graded R-module with $(C_*(X: R))_n = C_n(X: R)$. $C_n(X)$ will also be used to denote $C_n(X: Z)$. If A is a subcomplex of X, then the n-cells of X which are not cells of A will be referred to as the n-cells of X - A, and $C_n(X\text{-}A: R)$ will denote the free R-module generated by the n-cells of X - A. <u>Note</u>: $C_*(X\text{-}A)$ is not a chain complex.

We will assume, for the remainder of this section, that the topological group, G, is a C. W. complex, the identity, e, is a 0-cell of G, and the multiplication $G \times G \to G$ is a cellular map.

Lemma 3.5.2. If X is a C. W. complex, A is a subcomplex and the action $\phi: A \times G \to A$ is cellular, then $\overline{X}$ is a C. W. complex with subcomplexes $X \times e$ and A. The cells of $\overline{X} - A$ are in a 1-1 correspondence with the cells of $(X-A) \times G$. The cells of $\overline{X} - X$ are in a 1-1 correspondence with the cells of $(X-A) \times (G-e)$. The G action $\overline{X} \times G \to \overline{X}$ and the quotient map $X \times G \to \overline{X}$ are cellular.

Give $I = [0,1]$ the structure of a C. W. complex with one 1-cell (I) and two 0-cells (1) and (0). It follows from Lemma 3.5.1, that if X is a C. W. complex with 0-cell $x_o$, then $X \wedge I$ is a C. W. complex. If A is a subcomplex of X and $x_o \in A$, then $A \wedge I$ is a subcomplex of $X \wedge I$.

Lemma 3.5.3. Assume X is a C. W. complex, A is a subcomplex of X, $x_o$ is a 0-cell of X, and the contraction $h: A \wedge I \to A$ is cellular; then $\tilde{X} = (X \wedge I) \cup_h A$ can be given the structure of a C. W. complex such that the contraction $\tilde{h}: \tilde{X} \wedge I \to \tilde{X}$ and the quotient map $X \times I \to \tilde{X}$ are cellular. $X \wedge (1)$ is a subcomplex of $\tilde{X}$ and the cells of $\tilde{X} - (X \wedge 1)$ are in a 1-1 correspondence with the cells of $(X-A) \times (I-\dot{I})$.

It follows, from Lemmas 3.5.2 and 3.5.3, and by induction,

that each $D_n$ and each $E_n$ is a C. W. complex. The inclusions $D_n \hookrightarrow E_n$, $E_n \hookrightarrow D_{n+1}$ and the projections $(G \times I)^n \to D_n$, $(G \times I)^n \times G \to E_n$ are cellular. The contractions $h_n: D_n \wedge I \to D_n$ and the G-actions $\phi_n: E_n \times G \to E_n$ are also cellular. Thus $E_G$ is a C. W. complex, the contraction $h: E_G \wedge I \to E_G$ and the G-action $\phi: E_G \times G \to E_G$ are cellular maps. For $n \geq 1$ the cells of $D_n - E_{n-1}$ are in a 1-1 correspondence with n-tuples $[\sigma_1|\cdots|\sigma_n]$ of cells of $G - e$. The cells of $E_n - D_n$ are in a 1-1 correspondence with (n+1)-tuples $[\sigma_1|\cdots|\sigma_n]\sigma_{n+1}$ of cells of $G - e$ and the cells of $E_n - E_{n-1}$ are in a 1-1 correspondence with (n+1)-tuples $[\sigma_1|\cdots|\sigma_n]\sigma_{n+1}$ with $\sigma_i$ a cell of $(G-e)$ for $i = 1 \cdots n$ and $\sigma_{n+1}$ a cell of G. $[\ ]$ will denote the 0-cell of $D_o$ and if $\sigma$ is a cell of G, then $[\ ]\sigma$ will denote the corresponding cell of $E_o = G$. $[\sigma_1|\cdots|\sigma_n]$ denotes the cell in $D_n$ which is the image of $\sigma_1 \times (I) \times \sigma_2 \times (I) \times \cdots \times \sigma_n \times (I)$ under the projection $(G \times I)^n \to D_n$ and $[\sigma_1|\cdots|\sigma_n]\sigma_{n+1}$ denotes the cell in $E_n$ which is the image of $\sigma_1 \times (I) \times \sigma_2 \times (I) \times \cdots \times \sigma_n \times (I) \times \sigma_{n+1}$ under the projection $(G \times I)^n \times G \to E_n$. In particular $\dim([\sigma_1|\cdots|\sigma_n]) = (\sum_{i=1}^{n} \dim \sigma_i) + n$ and $\dim([\sigma_1|\cdots|\sigma_n]\sigma_{n+1}) = (\sum_{i=1}^{n+1} \dim \sigma_i) + n$ if $\sigma_i$ is a cell of $G - e$ for $i = 1 \cdots n$.

Next we investigate the cell structure on $B_G$ induced by the cell structure on $E_G$. For each $n \geq 0$, set $B_n$ equal to $E_n/G$.

Note $B_o = \{pt\}$. Recall that for each $n \geq 1$, $E_n - E_{n-1}$ is homeomorphic to $(D_n - E_{n-1}) \times G$. Thus for each $n \geq 1$ the projection $P: E_G \to B_G$ restricts to a relative homeomorphism $P: (D_n, E_{n-1}) \to (B_n, B_{n-1})$. It follows that each $n$, $B_n$ is a C. W. complex, the inclusions $B_n \to B_{n+1}$ and the projections $P: E_n \to B_n$ are cellular. For $n \geq 1$ the cells of $B_n - B_{n-1}$ are in a 1-1 correspondence with tuples $[\sigma_1|\cdots|\sigma_n]$ of cells of $G - e$, where $[\sigma_1|\cdots|\sigma_n]$ denotes the image under $p$ of the cell $[\sigma_1|\cdots|\sigma_n]$ in $D_n$. It can be seen that $B_G$ has the topology of the union, $\bigcup_{n \geq 0} B_n$. Hence $B_G$ is a C. W. complex and the map $P: E_G \to B_G$ is cellular. These results are summarized in the following theorem.

Theorem 3.5.4. Assume the topological group $G$ is a C. W. complex, the identity, $e$, is a 0-cell, and the multiplication $G \times G \to G$ is cellular. For $n \geq 0$ set $B_n$ equal to $E_n/G$. Then $E_G$ and $B_G$ are C. W. complexes and the map $p: E_G \to B_G$ is cellular. Each $E_n$ is a subcomplex of $E_G$ and each $B_n$ is a subcomplex of $B_G$. The projection $p: E_n \to B_n$ is cellular. For $n \geq 1$, the cells of $E_n - E_{n-1}$ are in a 1-1 correspondence with $n+1$-tuples, $[\sigma_1|\cdots|\sigma_n]\sigma_{n+1}$, of cells with $\sigma_i$ a cell of $G - e$ for $i = 1 \cdots n$ and $\sigma_{n+1}$ a cell of $G$, the cells of $B_n - B_{n-1}$ are in a 1-1 correspondence with $n$-tuples, $[\sigma_1|\cdots|\sigma_n]$, of cells of $G - e$.
$\dim([\sigma_1|\cdots|\sigma_n]\sigma_{n+1}) = (\sum_{i=1}^{n+1} \dim \sigma_i) + n$ and $\dim([\sigma_1|\cdots|\sigma_n]) = (\sum_{i=1}^{n} \dim \sigma_i) + n$. On the cell level, the map $p: E_G \to B_G$ is given

by $P([\sigma_1|\cdots|\sigma_n]) = [\sigma_1|\cdots|\sigma_n]$ and

$$P([\sigma_1|\cdots|\sigma_n]\sigma_{n+1}) = \begin{cases} [\sigma_1|\cdots|\sigma_n] & \text{if } \sigma_{n+1} \text{ is a 0-cell of } G \\ \text{degenerate if } \dim \sigma_{n+1} \geq 1 & \end{cases}.$$

It can be seen that if the topological group $G$ is abelian, then the multiplication in $B_G$ is cellular. Thus if $G$ is an abelian topological group, with cellular multiplication etc., then $B_G$ is an belian topological group, with cellular multiplication etc. So Milgram's Construction can be iterated in this case. Recall that a C. W. complex, X, is called an Eilenberg-MacLane space of type $(\pi,n)$ if

$$\pi_i(X) \cong \begin{cases} \pi & \text{if } i = n \\ 0 & \text{if } i \neq n \end{cases}.$$

Let $\pi$ be a group, and give $\pi$ the discrete topology. Then $\pi$ is an Eilenberg-MacLane space of type $(\pi,0)$, and $B_\pi$ is an Eilenberg-MacLane space of type $(\pi,1)$. Thus for any group $\pi$, there are Eilenberg-MacLane space of type $(\pi,0)$ and $(\pi,1)$. If $\pi$ is abelian, then $B_\pi$ is an abelian topological group with cellular multiplication. From repeated applications of Milgram's Construction, it follows that for any abelian group $\pi$ and any integer $n$, $n \geq 0$, there is an Eilenberg-MacLane space of type $(\pi,n)$ which is an abelian topological group with cellular multiplication. The significance of the Eilenberg-MacLane space $K$ of type $(\pi,n)$ is that for any C. W. complex $X$ there is a natural equivalence, $\eta$, of functors $\eta$: $[X,K]$ $H^n(X: \pi)$.

The rest of this section is concerned with showing $C_*(B_G: R)$ is isomorphic to the reduced bar construction applied to $C_*(G: R)$ and with calculating $H^*(B_{O(n)}: Z_2)$, $H^*(B_{U(n)}: R)$ and $H^*(B_{Sp(n)}: R)$.

Let $R$ be a commutative ring with unit.

Set $T^oC_*(G\text{-}e: R) = R$ for $i \geq 1$, set

$$T^iC_*(G\text{-}e: R) = \underbrace{C_*(G\text{-}e: R) \otimes_R \cdots \otimes_R C_*(G\text{-}e: R)}_{i}.$$

From Theorem 3.5.4, it follows that

$$C_*(E_G: R) \cong \bigoplus_{i=0}^{\infty} (T^iC_*(G\text{-}e: R)) \otimes_R C_*(G: R)$$

and

$$C_*(B_G: R) \cong \bigoplus_{i=0}^{\infty} (T^iC_*(G\text{-}e: R)).$$

Using $E_o = G$ and the cellular projections $E_{n-1} \times I \times G \to E_n$, it is possible to derive the formula for the differential induced on

$\bigoplus_{i=0}^{\infty} (T^iC_*(G\text{-}e: R)) \otimes_R C_*(G: R)$ by the differential on $C_*(E_G: R)$.

The formula is unnecessarily complicated by the fact that $C_*(G\text{-}e: R)$ is not a chain complex and by the fact that $C_*(G\text{-}e: R)$ is not closed under multiplication by cells in $G$.

These difficulties disappear after an appropriate normalization.

Before normalizing, we review the notion of graded algebra (following [Milnor and Moore]) and the definition of the reduced bar construction.

Recall: A graded R-module, A, is a family of R-modules $\{A_n\}$ indexed by the non-negative integers. A map of graded R-modules, $f: A \to B$, is a family $\{f_n\}$ of maps of R-modules, $f_n: A_n \to B_n$. The graded R-module, A, is called a Differential Graded R-module, DGR module for short, if for each $n \geq 1$ there is a map $d_n: A_n \to A_{n-1}$ of R-modules such that $d_n \circ d_{n+1} = 0$. The map $f: A \to B$ is called a map of DGR-modules if

$$\begin{array}{ccc} A_{n+1} & \xrightarrow{f_{n+1}} & B_{n+1} \\ \downarrow{\scriptstyle d_{n+1}} & & \downarrow{\scriptstyle d_{n+1}} \\ A_n & \xrightarrow{f_n} & B_n \end{array}$$

commutes for all $n \geq 0$.

The ring R will be considered as a DGR-module which is the 0-module in all dimensions except 0, and the ring R in dimension 0.

<u>Example</u>. For X a C. W. complex, $C_*(X: R)$ is a DGR-module. A cellular map $f: X \to Y$ induces a map $f_*: C_*(X: R) \to C_*(Y: R)$ of DGR-modules.

If A and B are graded R-modules, then $A \otimes_R B$ denotes the graded R-module with $(A \otimes_R B)_n = \bigoplus_{i+j=n} A_i \otimes_R B_j$. If

A and B are DGR modules, then $A \otimes_R B$ denotes the DGR module with differential defined by $d(a \otimes b) = da \otimes b + (-1)^{\dim a}\, a \otimes db$.

An algebra over R is a graded R-module, A, together with maps of graded R-modules, $\phi: A \otimes_R A \to A$ and $\eta: R \to A$ such that the diagrams

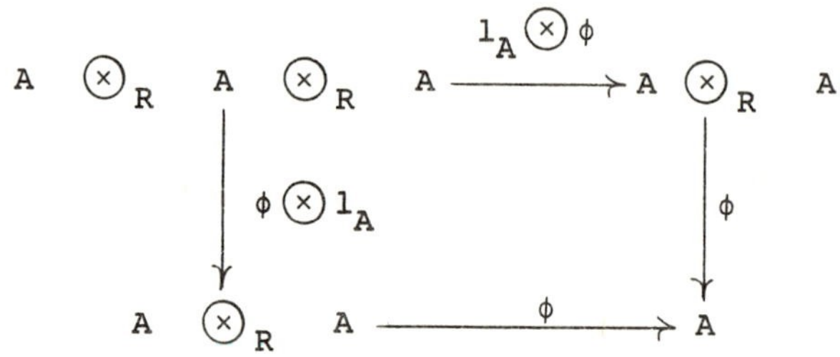


and

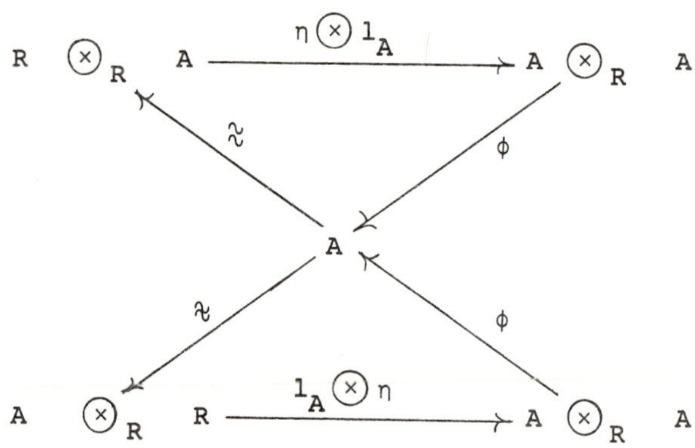


are commutative. $\phi$ is called the multiplication and $\eta$ is called the unit. If A is a DGR module and the maps $\phi$ and $\eta$ are maps of DGR modules, then A is called a DGR algebra.

Example. Let $G$ be a topological group and a C. W. complex such that the identity, $e$, is a 0-cell and the multiplication $u: G \times G \to G$ is cellular. Let $\eta: e \to G$ be the inclusion. Then $H_*(G: R)$ is a graded algebra over $R$ with multiplication $\phi_*: H_*(G: R) \otimes_R H_*(G: R) \to H_*(G: R)$ and unit $\eta_*: H_*(e: R) \cong R \to H_*(G: R)$. $C_*(G: R)$ is a DGR algebra with multiplication $\phi_*: C_*(G: R) \otimes C_*(G: R) \to C_*(G: R)$ and unit $\eta_*: C_*(e: R) \cong R \to C_*(G: R)$.

An augmentation of an algebra, $A$, is a map $\varepsilon: A \to R$ of algebras. The homomorphism $G \to e$ induces augmentations on $H_*(G: R)$ and $C_*(G: R)$.

Let $A$ be a DGR algebra with augmentation $\varepsilon: A \to R$. Then the reduced bar construction, $\overline{B}(A)$, is defined as follows.

Set

$$\overline{B}_0(A) = R.$$

For $i \geq 0$ set

$$\overline{B}_i(A) = \overline{A} \otimes_R \overline{A} \otimes_R \cdots \otimes_R \overline{A} \quad \text{where} \quad \overline{A} = \ker(\varepsilon: A \to R).$$

Set

$$\overline{B}(A) = \bigoplus_{i=0}^{\infty} \overline{B}_i(A).$$

The generator of $\overline{B}_0(A)$ is denoted by [ ] and a generator of $\overline{B}_i(A)$ for $i > 0$ is denoted by $[a_1|\cdots|a_i]$ where each $a_i$ is in $\overline{A}$. $\overline{B}(A)$ is a graded R-module with dim [ ] = 0 and $\dim[a_1 \cdots a_i]$ equal to $(\sum_{k=1}^{i} \dim a_k) + i$.

Note that $\bar{A}$ is a subchain complex of A, and $\bar{A}$ is a module over A, i.e. $\phi(\bar{A} \otimes A) \subseteq \bar{A}$. Define differentials $d_I, d_E$ and $\Delta$ on $\bar{B}(A)$ and $\bar{B}(A) \otimes_R A$ as follows.

1. $$d_I[a_1|\cdots|a_n]a_{n+1} = \sum_{i=1}^{n} [\bar{a}_1|\cdots|\bar{a}_{i-1}|da_i|\cdots|a_n]a_{n+1} + [\bar{a}_1|\cdots|\bar{a}_n]da_{n+1}$$

where $\bar{a} = (-1)^{1+\dim a}(a)$.

1'. $$d_I[a_1|\cdots|a_n] = \sum_{i=1}^{n} [\bar{a}_1|\cdots|\bar{a}_{i-1}|da_i|\cdots|a_n].$$

2. $$d_E[a_1|\cdots|a_n]a_{n+1} = \sum_{i=1}^{n-1} [\bar{a}_1|\cdots|\bar{a}_{i-1}|\bar{a}_i a_{i+1}|a_{i+2}|\cdots|a_n]a_{n+1} + [\bar{a}_1|\cdots|\bar{a}_{n-1}]\bar{a}_n a_{n+1}.$$

2'. $$d_E[a_1|\cdots|a_n] = \sum_{i=1}^{n-1} [\bar{a}_1|\cdots|\bar{a}_{i-1}|\bar{a}_i a_{i+1}|a_{i+2}|\cdots|a_n]$$

set $\Delta = d_I - d_E$ in both cases. Define $P_*: \bar{B}(A) \otimes_R A \to \bar{B}(A)$ by $p_*([\ ]) = [\ ]$ and $p_*([a_1|\cdots|a_n]a_{n+1}) = [a_1|\cdots|a_n]\ \varepsilon(a_{n+1})$. Then $\bar{B}(A) \otimes_R A$ with differential $\Delta$, and $\bar{B}(A)$ with differential $\Delta$ are DGR modules and $P_*$ is a map of DGR modules.

We are now ready to normalize. Let $\bar{C}_*(G\colon R)$ be the kernel of the augmentation induced on $C_*(G\colon R)$ by the continuous homomorphism $\varepsilon\colon G \to e$. Define

$$\psi\colon C_*(G\text{-}e\colon R) \to \overline{C}_*(G\colon R) \quad \text{by} \quad \psi(a) = a - \varepsilon_*(a)e.$$

Then $\psi$ is an isomorphism of graded R-modules.

Orient (I) the 1-cell of $I$ so that $d(I) = (1) - (0)$, and orient each 1-cell $\sigma$ of G, with vertices $\{a,e\}$, so that $d\sigma$ has the form $a - e$. Define $\psi\colon C_*(E_G\colon R) \to \overline{B}(C_*(G\colon R)) \otimes_R C_*(G\colon R)$ by $\psi([\ \ ]) = [\ \ ]$ and $\psi([\sigma_1|\cdots|\sigma_n]\sigma_{n+1}) = ([\psi\sigma_1|\psi\sigma_2|\cdots|\psi\sigma_n]\sigma_{n+1})$ and define $\psi\colon C_*(B_G\colon R) \to \overline{B}(C_*(G\colon R))$ by $\psi([\ \ ]) = [\ \ ]$ and $\psi([a_1|\cdots|a_n]) = [\psi a_1|\cdots|\psi a_n]$. The following theorem is proved by direct calculation along the lines indicated after Theorem 3.5.4.

<u>Theorem 3.5.5</u>. The diagram

$$\begin{array}{ccc}
C_*(E_G\colon R) & \xrightarrow{\quad\psi\quad} & \overline{B}(C_*(G\colon R)) \otimes_R C_*(G\colon R) \\
\Big\downarrow P_* & & \Big\downarrow P_* \\
C_*(B_G\colon R) & \xrightarrow{\quad\psi\quad} & \overline{B}(C_*(G\colon R))
\end{array}$$

is a commutative diagram of DGR modules, where the differential on $\overline{B}(C_*(G\colon R)) \otimes_R C_*(G\colon R)$ and on $\overline{B}C_*(G\colon R)$ is $\Delta = d_I - d_E$. The maps $\psi$ are isomorphisms of DGR modules. Thus

$$\psi_*\colon H_*(B_G\colon R) \to H(\overline{B}(C_*(G\colon R)),\ \Delta = d_I - d_E)$$

is an isomorphism.

Remarks.

1. $\overline{B}(C_*(G: R)) \otimes_R C_*(G: R)$ is a proper projective resolution of $R$ as a $C_*(G: R)$ module, see [Smith 1]. Thus $H(\overline{B}(C_*(G: R),\Delta)$ is denoted by $D\ \mathrm{Tor}_{C_*(G:\ R)}(R,R)$.

2. For an arbitrary DGR algebra $A$, it is not likely that $H(\overline{B}(A),\Delta)$ will yield to calculation by brute force.

3. Neglecting the differential on $C_*(G: R)$, $\overline{B}(C_*(G: R)) \otimes_R C_*(G: R)$ with differential $d_E$ is a projective resolution of $R$ as a $C_*(G: R)$ module. Thus $H(\overline{B}(C_*(G: R),d_E)$ is denoted by $\mathrm{Tor}_{C_*(G:\ R)}(R,R)$.

4. If the differential on $C_*(G: R)$ is zero then $H_*(B_G: R)$ is isomorphic to $\mathrm{Tor}_{H_*(G:\ R)}(R,R)$. Thus $H_*(B_{O(n)}: Z_2)$ $\cong \mathrm{Tor}_{H_*(O(n):\ Z_2)}(Z_2,Z_2)$ and $H_*(B_{G(n)}: R)$ $\cong \mathrm{Tor}_{H_*(G(n):\ R)}(R,R)$ for $d = 2$ or $4$ and any commutative ring $R$. (See sections 2 and 3 of Chapter 1).

In order to adequately describe $\mathrm{Tor}_{H_*(G(n):\ R)}(R,R)$ it is necessary to discuss coalgebras.

Definition. A coalgebra over $R$ is a graded R-module, $A$, together with maps of graded R-modules

$$\phi: A \to A \otimes_R A \quad \text{and} \quad \varepsilon: A \to R$$

such that the diagrams

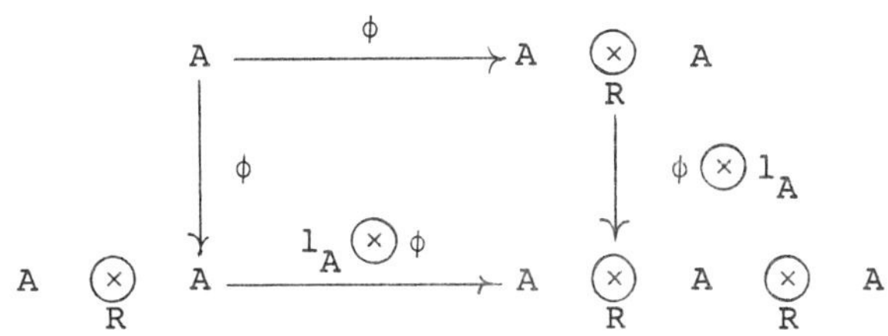

and

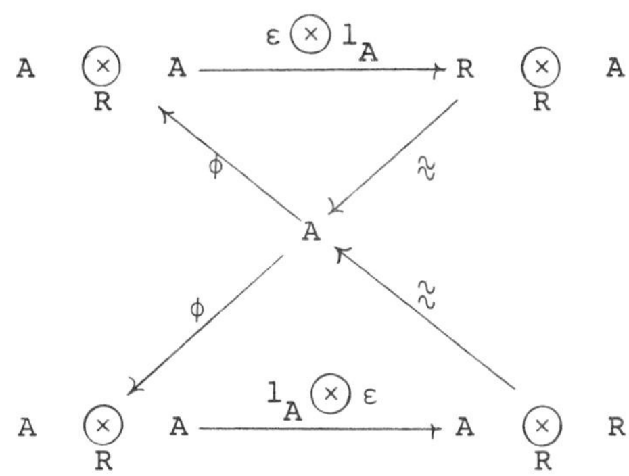

commute. If $A$ is a DGR module and the maps $\phi$ and $\varepsilon$ are maps of DGR modules, then $A$ is called a DGR coalgebra.

<u>Examples</u>.

1. $\Gamma(x)$ = divided polynomial coalgebra on one generator $x$ of dimension $k$, with $k \geq 1$.

$$(\Gamma(x))_n = \begin{cases} R \text{ generated by } \delta_o(x) & n = 0 \\ R \text{ generated by } \delta_j(x) & n = jk \\ 0 \text{ otherwise} \end{cases}$$

$\varepsilon: \Gamma(x) \to R$ is defined by $\varepsilon(\delta_o(x)) = 1$

$$\varepsilon(\delta_j(x)) = 0 \quad j \geq 1$$

$\phi: \Gamma(x) \to \Gamma(x) \otimes_R \Gamma(x)$ is defined by

$$\phi(\delta_n(x)) = \sum_{p+q=n} \delta_p(x) \otimes \delta_q(x) \quad \text{for } n \geq 0.$$

2. $\overline{B}(A)$. Define $\varepsilon: \overline{B}(A) \to R$ by $\varepsilon([\ \ ]) = 1$

$$\varepsilon([a_1|\cdots|a_n]) = 0 \quad \text{for } n \geq 0.$$

Define $\phi: \overline{B}(A) \to \overline{B}(A) \otimes \overline{B}(A)$ by

$$\phi([\ \ ]) = [\ \ ] \times [\ \ ]$$

and

$$\phi([a_1|\cdots|a_n]) = \sum_{i=1}^{n-1} [a_1|\cdots|a_i] \otimes [a_{i+1}|\cdots|a_n]$$

$$+ [\ \ ] \otimes [a_1|\cdots|a_n] + [a_1|\cdots|a_n] \otimes [\ \ ].$$

Then $\overline{B}(A)$ is a DGR coalgebra.

3. If $A$ and $B$ are coalgebras over $R$, define $\phi: (A \otimes_R B)$ $(A \otimes_R B) \otimes_R (A \otimes_R B)$ to be the composition

$$A \underset{R}{\otimes} B \xrightarrow{\phi_A \otimes \phi_B} (A \underset{R}{\otimes} A) \underset{R}{\otimes} (B \underset{R}{\otimes} B)$$

$$\xrightarrow{1_A \otimes T \otimes 1_B} (A \underset{R}{\otimes} B) \underset{R}{\otimes} (A \underset{R}{\otimes} B)$$

where $T(a \otimes b) = (-1)^{(\dim a)(\dim b)} (b \otimes a)$.

Set $\varepsilon: A \otimes_R B \to R$ equal to the composition $A \otimes_R B \xrightarrow{\varepsilon_A \otimes \varepsilon_B} R \otimes_R R \approx R$. Then $(A \otimes B)$ is a co-algebra over $R$. Denote $\Gamma(x_1) \otimes_R \Gamma(x_2) \otimes_R \cdots \otimes_R \Gamma(x_n)$ by $\Gamma(x_1, \ldots x_n)$.

4. If $H_i(X: R)$ is a projective $R$-module for all $i$ then $H_*(X: R)$ is a coalgebra over $R$, with $\phi: H_*(X: R) \to H_*(X: R) \otimes_R (X: R)$ induced by the diagonal map $X \to X \times X$ $(x \to (x,x))$ and $\varepsilon: H_*(X: R) \to R$ induced by the map $X \to pt$.

From [Milgram] it follows that

Theorem 3.5.6. The diagram

$$\begin{array}{ccc} H_*(B_G: R) & \xrightarrow{(D)_*} & H_*(B_G \times B_G: R) \\ \downarrow \psi_* & & \downarrow (\psi \otimes \psi)_* \\ H(\overline{B}C_*(G: R) & \xrightarrow{\phi_*} & H(\overline{B}C_*(G: R) \underset{R}{\otimes} \overline{B}C_*(G: R)) \end{array}$$

commutes where $D: B_G \to B_G \times B_G$ is the map $x \to (x,x)$ and $\phi$ is the map defined in example 2 above.

By direct calculation (see for example [Smith 2], pp. 71-72), it follows that

$$\mathrm{Tor}_{H_*(O(n):\ Z_2)}(Z_2,Z_2) \cong \Gamma([1],[2],\ldots.[n] \quad \text{where} \quad \dim[i] = i.$$

and

$$\mathrm{Tor}_{H_*(G(n):\ R)}(R,R) \cong \Gamma([1],[2],\ldots.[n]) \quad \text{where} \quad \dim[i] = id$$
$$\text{for} \quad d = 2 \text{ or } 4,$$

and the isomorphisms above are isomorphisms of coalgebras.

If $A$ is a graded R-module, define the graded R-module, $A^*$, by setting $(A^*)_n = \mathrm{Hom}_R(A,R)$. Note that a map $f: A \to B$ of graded R-modules induces a map of graded R-modules $f^*: B^* \to A^*$. If $A$ is a coalgebra over $R$, then $A^*$ is an algebra over $R$ with multiplication

$$A^* \underset{R}{\otimes} A^* \to (A \underset{R}{\otimes} A)^* \xrightarrow{\phi^*} A^* \text{ and unit } R \xrightarrow{\varepsilon^*} A^*$$

where

$$A^* \underset{R}{\otimes} A^* \to (A \underset{R}{\otimes} A)^*$$

is defined by

$$f \otimes g(a \otimes b) = f(a)g(b).$$

Recall that if each module $H_i(X: R)$ is a finitely generated free R-module, then $(H_*(X: R))^*$ is isomorphic, as a graded R algebra, to the cohomology ring, $H^*(X: R)$ and $(\Gamma(x_1 \cdots x_n))^* \cong P(x_1^*, \ldots x_n^*)$, the polynomial algebra over R generated by the classes dual to the $x_i$, $i = 1 \cdots n$.

These results are summarized in

<u>Theorem 3.5.7</u>. For $n \geq 1$, $H^*(B_{O(n)}: Z_2) \cong P(w_1, \ldots . w_n)$ as algebras over $Z_2$, where $\dim w_i = i$.

For $n \geq 1$, R any commutative ring with unit, $H^*(B_{U(n)}: R) \cong P(c_1, \ldots c_n)$ as algebras over R, where $\dim c_i = 2i$.

For $n \geq 1$, R any commutative ring with unit, $H^*(B_{Sp(n)}: R) \cong P(p_1, \ldots p_n)$ as algebras over R, where $\dim P_i = 4i$.

## BIBLIOGRAPHY

Richard Arens; Topologies for Homeomorphism Groups, Amer. Jour. Math. 68(1946) pp. 593-599.

A. Borel; Topology of Lie Groups and Characteristic Classes, (Survey) Bull. Amer. Math. Soc. 6(1955) pp. 397-432.

Claude Chevalley; Theory of Lie Groups I, Princeton Mathematical Series No. 8, Princeton University Press, 1946.

A. Dold and R. Lashoff; Principal Quasifibrations and Fibre Homotopy Equivalence of Bundles, Illinois J. Math. 3(1959) pp. 285-305.

F. Hirzenbruch; Topological Methods in Algebraic Geometry, (3rd edition translated), J. Springer, 1966.

Dale Husemoller; Fibre Bundles, McGraw-Hill series in Higher Mathematics, 1966.

S. MacLane; Homology, J. Springer, 1963, Grundlehren Series No. 114.

W. S. Massey; Algebraic Topology: An Introduction, Harbrace College Mathematics Series, Harcourt, Brace and World, Inc., 1967.

R. J. Milgram; The Bar Construction and Abelian H-Spaces, Illinois Jour. Math. 11(1967) pp. 242-250.

C. E. Miller; The Topology of Rotation Groups, Ann. of Math. 57(1953) pp. 90-113.

J. Milnor; Lectures on Characteristic Classes, Annals of Mathematics Studies No. 76, Princeton University Press, 1973.

________; Construction of Universal Bundles II, Ann. of Math. (2) 63(1956) pp. 430-436.

J. Milnor and J. Moore; On the Structure of Hopf Algebras, Ann. of Math. (2) 81(1965) pp. 211-264.

J. Moore; Algèbre homologique et homologie des espaces classifiants; exposé 7 in Séminaire Henri Cartan $12^e$ annee: 1959/60.

Larry Smith; Homological Algebra and the Eilenberg-Moore Spectral Sequence, Trans. Amer. Math. Soc. 129(1967) pp. 58-93.

__________; Lectures on the Eilenberg-Moore Spectral Sequence, Springer-Verlag Lecture Notes in Mathematics, Vol. 134, 1970.

E. H. Spanier; Algebraic Topology, McGraw-Hill Series in Higher Mathematics, McGraw-Hill, 1966.

N. E. Steenrod; Cohomology Operations, Lectures by N. E. Steenrod written and revised by D. B. A. Epstein, Annals of Mathematics Studies No. 50, Princeton University Press, 1962.

____________; The Topology of Fibre Bundles, Princeton Mathematical Series No. 14, Princeton University Press, 1951.

____________; Milgram's Classifying Space of a Topological Group, Topology 7(1968) pp. 349-368.

____________; A Convenient Categroy of Topological Spaces, Mich. Math. J. 14(1967) pp. 133-152.

J. H. C. Whitehead; Combinatorial Homotopy, Bull. Amer. Math. Soc. 55(1949) pp. 213-243.

I. Yokota; On the Cellular Decompositions of Unitary Groups, J. Inat. Polytech. Osaka City University 7(1956) pp. 39-49.

INDEX

U



W